SEVEN
WORLDS
ONE
PLANET

SEVEN WORLDS ONE PLANET

NATURAL WONDERS FROM EVERY CONTINENT

JONNY KEELING and **SCOTT ALEXANDER**

Foreword by **DAVID ATTENBOROUGH**

BBC
BOOKS

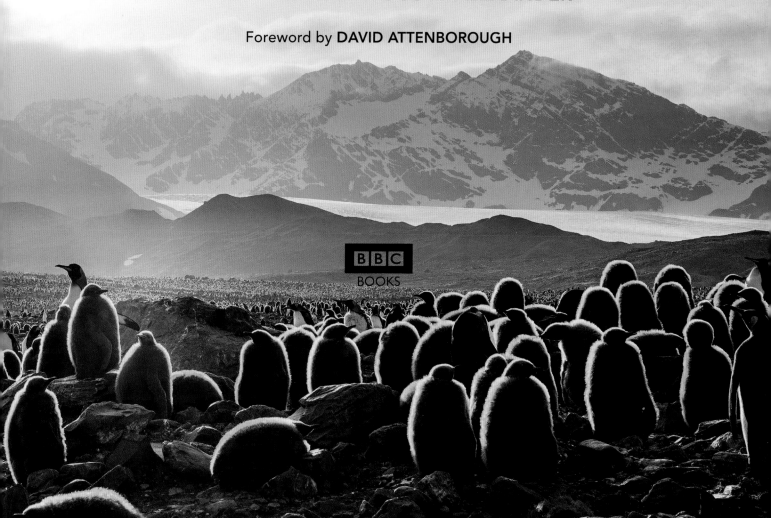

Contents

Foreword

Why don't polar bears eat penguins? The riddle is probably as old as any other that you will find in a Christmas cracker. And the answer is not difficult to work out: penguins and polar bears live at opposite ends of the world and never meet.

But why is that so? To answer that question, you have to consider the Earth's geological history. Three hundred million years ago, the only land on this planet's otherwise ocean-covered surface was a single super-continent. It was there that terrestrial life began. Eventually, however, this immense landmass began to break up. One fragment started to drift south. As it approached the pole, it became so cold that none of its animal passengers were able to survive. This was the continent we now call Antarctica and no land animals – except human beings – have ever managed to reach it since.

Questions about why different animals live where they do are likely to occur to anyone who watches a natural history series surveying the entire globe as does the one on which this book is based. But not all are so easily answered. Why, for example, is it deer that nibble grass in North America, whereas the medium-sized mammals that live in Africa in a similar fashion and with a similar diet are antelope? Or why are there apes in the tropical forests of Africa and Asia but none at all in the jungles of South America?

Below and Overleaf
Sir David Attenborough and the stark beauty of Vestrahorn on Iceland's Stokksnes peninsula. The sea has eroded the dark rocks, the sediments washed ashore to form black sand beaches and dunes.

The introduction that follows these pages helps to answer such questions. It also explains why communities of animals and plants on the seven continents of our planet are still so different from one another that they can justifiably be described as separate worlds.

Each has its own particular animal treasures. Some are rare and little known. Take, for example, the olm that lives only in the caves of eastern Europe. It is a kind of salamander, as long and as slim as a small snake, and has two pairs of diminutive legs and a moist, scale-less skin. Because it lives in permanent darkness, it has lost its eyes and the pigment in its skin and so has become a ghostly white. Now its life is so uneventful and requiring so little energy that it only needs to eat once in a decade.

Or consider the blue-faced golden-coated monkey that lives in northern China in places that are snow-covered for at least five months of the year. It is so cold there that the monkeys have developed thick furry coats and reduced the danger of their noses being frostbitten by evolving ones that are so severely snubbed that they can hardly be described as noses at all. Both these strange creatures are rare and scarcely known because they live in restricted and little-visited habitats.

There are, however, other rarities whose numbers are also small but for very different and more alarming reasons. They were once abundant but we have displaced them from the territories that were once theirs. Sometimes we

Below
Sir David is with one of only two northern white rhinos, both female, which survive in the wild in Kenya's Ol Pejeta Conservancy. The last male died in 2018, rendering the subspecies effectively extinct.

have done so for the most trivial of reasons. During the nineteenth century, European settlers both in Australia and North America introduced blackbirds and thrushes from Europe because they considered that their songs were more melodious than those of the local birds. They imported foxes because the hunters among them had nothing they thought suitable for the chase. And some brought their pet cats because they enjoyed having them sitting purring by the fireside. Some of these introductions failed and died out after a few generations. But others flourished and became plagues that had catastrophic effects on the indigenous animal populations.

Others of our introductions have been accidental rather than deliberate, as on the many occasions when we have allowed rats, hitchhiking on our ships, to escape ashore in territory where they never existed before. Again and again these hardy, omnivorous and prolific intruders have then caused havoc among the local animals which had no defence against them.

But the greatest changes we have caused are those we have made deliberately in order to provide for our ever-increasing numbers. We have felled forests, drained swamps, and covered fertile meadows with concrete in order to build our homes and factories, airports and motorways. So, over the past two hundred years, wild animals that were once relatively abundant on every continent except Antarctica have been decimated and now survive in dangerously small numbers. Such are the Iberian lynx, the European wolf, the Tasmanian tiger, the Arctic musk ox and many others.

Seven Worlds, One Planet describes and illustrates rarities of both kinds, together with some of the most dramatic natural wonders that still remain on Earth. Let us hope that our increasing understanding of the way nature functions will at last persuade people everywhere to care for the animals that evolved on this planet's continents and allow them the space they need to live in the particular world that once was theirs.

David Attenborough

Introduction

About 200 million years ago, our planet looked very different from the Earth we see today. There was only one continent, the supercontinent Pangaea, surrounded by a single superocean Panthalassa, but then something monumental happened. Gargantuan forces began to rip apart the Earth's crust and Pangaea broke into two immense mega-continents, which then drifted apart. Millions of years later, these split again, and then again and again, eventually each part separating from its neighbours and moving across the face of the Earth to form the seven great continents we see today.

Riding on and around them were the descendants of plants and animals that survived the many mass extinction events that have occurred during our planet's history. With the break-up of Pangaea, however, came new land, new seas, and new habitats. Plants and animals adapted to their new homes. Groups of organisms became separated from their original populations and evolved into new species. Life proliferated and diversified, but that's not all. The changing position of the continents has had a huge effect on ocean currents, atmospheric circulation and the strength of the seasons. The splitting up of Pangaea and the formation and movement of the continents shook up the world, and its effects are still being felt today.

FLOATING CONTINENTS

That the continents move across the surface of the Earth was widely recognised in the late 1960s, but the idea was first proposed long ago. In 1596, Antwerp-born geographer and mapmaker Abraham Ortelius put forward the idea that the Americas were 'torn away from Europe and Africa … by earthquakes and floods', and he suggested that one could see the 'vestiges of the rupture' if one looked at the coasts of the three continents on a world map. Since then, many other distinguished observers – Alexander von Humboldt and Alfred Russel Wallace among them – have remarked on how, for instance, the east coast of South America fits snugly with the west coast of Africa, but it was not until 1912 that the German geophysicist and meteorologist Alfred Wegener, independently of his predecessors, came up with the view that the continents moved across the face of the Earth. He called it quite simply *Kontinentalverschiebung*, meaning 'continental drift'.

Below
Fossils of the freshwater reptile *Mesosaurus* have been found on either side of the Atlantic Ocean, indicating South America and Africa were once joined.

Opposite
Like a scene from Earth's ancient past, molten lava pours into the ocean on the coast of Maui, Hawaii.

At first, a highly sceptical scientific community greeted his ideas, but in the 1950s, the evidence began to build. Palaeontologists found the same early mammal fossils in South Africa, India and Antarctica, so those locations must have been closer together in the past. Similarly, the fossils of a small freshwater reptile were found in South America and Africa on either side of the Atlantic, an indication that an ocean had not always separated the two continents. Geologists, meanwhile, revealed that the Appalachian Mountains on North America's east coast have a close geological affiliation with the Caledonian Mountains in Scotland, and biologists pointed to the occurrence of living marsupials in South America and Australasia, with fossil species in Antarctica.

Yet, even with such evidence, many scientists remained doubtful. 'If the continents drift about,' they asked, 'what causes them to move?' English geologist Arthur Holmes, however, had already provided an answer. In 1930, Holmes showed how convection currents in the Earth's mantle could be the mechanism that propels the continents across its surface. It all started a billion or so years after the Earth had formed about 4.54 billion years ago.

SHIFTING PLATES

Directly after our planet's birth, its surface was covered with oceans of molten magma. As the Earth lost heat and cooled, it began to form the onionskin-like structure with which we are familiar today: a solid crust, a viscous mantle, a liquid outer core and a solid inner core. All those billions of years ago, however, as parts of the crust and upper mantle – known as the lithosphere – cooled, they became denser and began to sink into the lower mantle, where they were warmed up again from the heat left over from the planet's birth and the decay of radioactive elements. Convection currents did the rest.

Holmes suggested that magma returns to the surface when heat from the Earth's interior causes it to well up, like boiling porridge, and ooze out at mid-ocean ridges and other spreading centres, such as the East African Rift. At the surface, the magma cools and is pulled slowly across the Earth's surface towards the other side of the plate, like a tablecloth sliding off a table. Here, it is dragged below the neighbouring plate, known as a subduction zone, and sinks back down into the mantle and is recycled.

It gave rise to the theory of plate tectonics, which proposes that the Earth's lithosphere is divided into large and small plates, like a gigantic moving jigsaw puzzle, with plates always being formed or destroyed. It means that the Earth beneath our feet is not permanent and unmoving: it is continually in motion, and constantly recycled and renewed, in some places faster than in others.

The most geologically active regions are at the plate margins, where plates collide, scrape together, move apart, or are pushed one below the other. They are the places where violent earthquakes and volcanic eruptions occur most frequently, and where the Earth's crust buckles into great mountain chains or sinks to form deep ocean trenches. Way back in prehistory, it was these same geological forces that caused Pangaea to break up.

Opposite
Silfra Canyon, in Iceland's Thingvellir National Park, is a rift in the Earth's crust forming the boundary between the American plate on the left and the Eurasian plate on the right.

NEW WORLD ORDER

Pangaea split first into Laurasia in the north, which included what is now most of Europe, Asia and North America, and in the south Gondwana, with Africa, South America, India, Antarctica and Australasia.

A North Atlantic–Caribbean rift, forerunner of the North Atlantic Ocean and the Caribbean Sea, separated Laurasia from South America and Africa. At about the same time, Gondwana began to split apart. A rift, that would become the Indian Ocean, separated South America and Africa from India, Australasia and Antarctica. Some time later, India peeled away from the other two.

By about 140 million years ago, a new rift had opened up between South America and Africa. This would become the South Atlantic Ocean. Not long after, Spain rotated clockwise, forming the Bay of Biscay, and collided with France, pushing up the Pyrenees. Australia separated from Antarctica, and Madagascar split away from Africa. About 90 million years ago, India parted company with Madagascar, and both India and Africa were moving independently northwards.

By 65 million years ago, Australia was heading northwards too, while New Zealand split away from its east coast and Antarctica twisted and moved to the south and west, eventually settling at the South Pole. At the same time, the North and South Atlantic oceans widened. Africa was brought to a halt by Europe and rotated slightly clockwise, trapping the remnants of an ancient sea to form the Mediterranean and pushing up the Alps and Atlas Mountains. India collided with Asia, building the Himalayas. By about 20 million years ago, the continents were not far from the positions we see them today.

STILL MOVING

The movements of the continents, of course, have not stopped. Magma continues to ooze from the Mid-Atlantic Ridge, creating new ocean floor on either side, so North America and Europe are moving away from each other by about 2.5 centimetres a year. India continues to push against Eurasia, buckling the land, so the Himalayas are still growing. Nanga Parbat, the so-called 'killer mountain' and the ninth-highest mountain in the world, for example, is growing taller by 7 millimetres a year, the fastest-growing mountain on the planet. Even more startling is Australasia. It is heading rapidly, at about 7 centimetres a year, for a collision with Southeast Asia; in fact, it is moving so fast GPS cannot keep up with it. The last calibration was made in 1994, when the island continent was close to 200 metres further north than on a previous correction; since then, it has travelled northwards for another 1.5 metres and rotated slightly clockwise. Eventually, it will smash into Southeast Asia, creating another extensive mountain chain.

PLATE TECTONICS AND BIODIVERSITY

As well as defining the shape and position of the continents, creating basic landscapes and seascapes, and contributing significantly towards each

Above
The supercontinent Pangaea (top) broke first into two mega-continents (middle), and then into the seven continents we see today (bottom).

continent's climate, the movement of the tectonic plates has had an influence on the living world, especially on biodiversity. When Pangaea fused together, on went the evolutionary brakes. Biodiversity stabilised and maybe even declined because there was little environmental pressure for a species to change. When the supercontinent broke up and the land fragmented, things sped up dramatically. Terrestrial organisms were isolated on the separating landmasses and were moved into locations with new and maybe challenging climates. Marine animals – especially those in shallow waters bordering continents – were carried to other parts of the globe. As both became adapted to the changing conditions in their new habitats, many evolved into new species.

The ocean itself was also in a state of flux. At the time Pangaea broke up, sea levels were rising, and by the late Cretaceous, they were up to 250 metres higher than they are today. The sea inundated lowland areas on the new continents, causing more fragmentation and more speciation. Europe, for example, was divided up into many small islands, each with its own evolving species, so biodiversity across Europe was far higher than on a single landmass of equivalent size.

Without the break-up of Pangaea and the continued evolution of our continents, it is clear that the world would be a biologically poorer place. When they became isolated in their respective parts of the planet, each continent developed its own terrain and climate, and evolved and continues to evolve its own flora and fauna, giving rise to the great diversity of life on the Earth today.

Below
Nanga Parbat in the Himalayas is the world's fastest-growing mountain.

SEVEN WORLDS

Australasia may be whizzing across the globe at high speed, but, because it is in the middle of the Australian plate, there has been little tectonic upheaval and few major geological changes for at least three million years. Without widespread mountain building, the land has had long exposure to the forces of erosion, so the entire island continent is relatively flat and very old rocks are exposed. The absence of large mountain chains, aside from the Australian Alps in the southeast, has also influenced the climate, especially the lack of rain or snow usually produced when moist air moves over mountains, but that's not the only contribution to Australia's arid centre. The main rain-bearing winds sweep in from the southwest, having blown over cold ocean currents, so there is less evaporation from the ocean and therefore poor cloud formation. Rain also decreases the further away from the coast you are, which means very little rain reaches the interior, making it the second-driest continent in the world, after Antarctica. Rain, and lots of it, falls in the northeast, however, influenced by the monsoon and the proximity of the Pacific Ocean, giving rise to tropical rainforests, probably the oldest on Earth.

Above
The limestone formations of Pinnacles Desert, Nambung National Park, are in the dry centre of Western Australia.

Africa is basically a huge plateau made from five very old and stable parts of the lithosphere that were formed about 2.8 billion years ago – that's only 1.75 billion years after the Earth was formed. In more recent times, after the break-up of the southern mega-continent of Gondwana, Africa moved slowly northwards until it crashed into Europe and, straddling the Equator, ground to a slow crawl, so its ancient rocks have barely moved across the globe for many millions of years. It did, however, park itself over a gigantic magma plume, probably the same bulge that caused Pangaea to break apart. This has caused the crust to split and the two halves to separate, forming the East African Rift, which is tearing East Africa apart from Ethiopia to Mozambique, and might one day give rise to a new ocean where the Red Sea is today. Because of its position on the globe, Africa has more hot deserts than any other continent. The Sahara is the world's biggest, and the Namib one of its oldest. While the Sahara is inland, and has not always been a desert, the Namib abuts the Atlantic Ocean, yet it has been dry or semi-dry for 55–80 million years. This is because of the cold ocean currents that move along the coast. They cool the air so it holds less moisture, and any rain falls before the clouds reach land. Instead, thick fogs envelop the dunes, so life can still survive here.

Above
Fog rolls in from the sea and envelops the sand dunes of the Namib Desert.

Europe was an island continent about 34 million years ago. It was separated from Asia by a shallow sea but, as sea levels changed, sediments filled the gap so Europe became the westernmost part of the combined continents of Eurasia. At this time, animals invaded from Asia, replacing many of Europe's animals in a local mass extinction event known as the 'Grande Coupure'. Europe's prehistory from then on was dominated by the coming and going of Ice Ages, periods of intense cold, when ice sheets covered northern Europe, interspersed with warm periods, when hippopotamuses were in Trafalgar Square. When the ice melted after the last Ice Age, sea levels rose, Britain's continental shelf was flooded, and so the place became the British Isles. Almost all of northern Europe's fauna and fauna, including modern humans, moved in after the ice had disappeared. Today, the close proximity of the Atlantic Ocean and the movements of warm waters in the Gulf Stream and associated currents dominate the climate, especially of Western Europe. It means the climate is not extreme, but generally mild and wet, with ice-free coasts even to the north of the Arctic Circle.

Asia has the lion's share of Eurasia, and is still growing, both upwards and outwards. The volcanoes and geyser fields of the remote Kamchatka Peninsula and the island arcs of Indonesia, the Philippines and Japan are regions where subduction of the floor of the Pacific Ocean has resulted in extreme and often violent volcanic and seismic activity. The rest of the continent is so vast that

Above
Exposed to the full fury of the Atlantic Ocean at the western edge of Europe, Stac a' Phris is a natural sea arch eroded by the waves on the Isle of Lewis, Scotland.

it extends across eleven time zones, and has a bit of everything. In the north, the tundra abuts the Arctic Ocean, with polar bears and walruses. Further south are the great coniferous forests of the taiga, with the largest subspecies of tiger. Deciduous forests follow, and then steppe, desert, and towering mountains like the Himalayas. South again, there are tropical forests and mangroves that lead to the coast and paradise islands, with coral reefs and exotic locations like the Coral Triangle. It is the most populous continent, but people are concentrated in dense conurbations, so there are vast unpopulated regions where wildlife can thrive relatively undisturbed.

North America's central core is composed of rocks over 2 billion years old. Across the mainland, younger sedimentary rocks overlay it, but some of the oldest rocks on Earth can be found at the surface on Greenland and in eastern Canada. Before Pangaea, an island chain bumped into the east coast pushing up the Appalachians. They were once as grand as the Himalayas, but the forces of erosion have since ground them down. Some time later, in the west, the Rockies were thrust up, giving the continent its two main north–south orientated mountain chains. With rising and falling sea levels, due to changing world climates, North America has been joined to and separated from both Asia and South America several times, so its fauna is a mix of native animals, such as the pronghorn and caribou, with bison, wolves and brown bears from Asia, and possums and porcupines from South America. It also

exported species. Members of the camel family originated in North America, but camels went east into Asia, and their llama-like relatives headed south to South America. Following them the length and breadth of the Americas was the puma, found today from Alaska to Terra del Fuego.

South America has kept itself very much to itself since the break-up of Pangaea and the splitting of the continents. Two landforms dominate: ancient rocks in the north and east, and the younger Andes mountain chain in the west. Ancient rocks can be seen in the north of the continent as sheer-sided mountains, known as tepuis, while the Andes on the Pacific coast are the result of the subduction of the Nazca Plate below the edge of the South American Plate, forming a volcanic and seismically active region, part of the 'Pacific Ring of Fire'. The Amazon river once flowed westwards into the Pacific but, when the Andes arose, the Amazon basin first became a shallow lake, and eventually its waters broke through the ancient mountains in the east to flow eastwards into the Atlantic. The continent has been joined to and separated from North America, from time to time, with an interchange of plants and animals each time a land bridge formed. Today, it is probably the richest continent, with

Above
Heavy seasonal rain falling in the eastern Andes and the northwest of the Amazon river basin, causes rivers to burst their banks and flood the rainforest on either side.

Opposite
Caribou or reindeer have lived in North America for at least 2 million years. They crossed eastwards across the Bering Land Bridge and spread throughout Eurasia.

40 per cent of the world's species found on just 12 per cent of the Earth's land surface. Most live in the cloud forests and tropical rainforests. A beetle expert revealed just how rich these forests must be. He identified more than 1,000 new species of beetles in a tropical rainforest tree, compared to the 50 known species found in an entire British oak wood.

Antarctica's fate was sealed when the land bridge between South America and Antarctica was breached, enabling the Antarctic Circumpolar Current to flow clockwise around the entire continent. The current, which dwarfs the Gulf Stream in the volume of water moved, effectively cuts off Antarctica from the rest of the world, reducing the southward flow of heat from the other oceans and causing widespread glaciation on the Antarctic continent. Antarctica was warmer in the distant past, but permanent ice caps formed when atmospheric CO^2 reduced significantly about 34 million years ago. Today, almost the entire mainland is covered by ice that has an average thickness of nearly two kilometres. It locks up about 70 per cent of all the freshwater on the planet. Today, there are two ice sheets – the West Antarctic and East Antarctic ice sheets, separated by the Transantarctic Mountains, and below the ice are more than 70 subglacial lakes, Lake Vostok being the largest. The ice is so heavy that the land beneath it has been pushed down about 500 metres into the Earth's crust, which means the continental shelf is deeper than those around other continents and plays host to deep-sea marine life. The ice sheets are melting, however, due to the warming planet, and the Antarctic mainland is actually rising at about 42 millimetres a year, four or five times faster than scientists had expected.

Above
Antarctica's Lemaire Channel is an often calm but sometimes iceberg-filled passage between Graham Land on the Antarctic Peninsula and Booth Island in the Wilhelm Archipelago.

SUPERCONTINENTAL CYCLE

Pangaea wasn't the first supercontinent during Earth's history, and it won't be the last. Supercontinents form every 300–500 million years. In the latest cycle, you can see the continents are already coming together. This kind of geological activity is accompanied by a natural decline in biodiversity as natural systems begin to stabilise, just as it did when Pangaea was formed.

Africa's collision with Europe and the pending demise of the Mediterranean Sea, India's crash with Asia, and Australia's imminent head-on collision with Southeast Asia, for instance, means that Eurasia will get a whole lot bigger, but there will be fewer shallow-water marine habitats around the enlarged continent than there were before these continental pile-ups.

Off the coast of Portugal in the eastern Atlantic, scientists are also seeing something quite remarkable. They are observing what they believe to be the start of one tectonic plate being pushed below another. This interpretation is controversial, but if it's correct the North Atlantic will start to shrink, rather than widen. What will be its impact on wildlife? One big difference these days is that geological events such as these are not necessarily the main driving force of biodiversity that they once were. Now, we have to factor in humans to the equation. We are rewriting nature's rulebook… and it's not for the better.

If the global statistics are correct, the planet is experiencing the worst period of plant and animal species dying out since the demise of the non-avian dinosaurs 66 million years ago; and, for every species that dies out, there are many others that depend on it, so they are threatened too. It means the number of extinctions will snowball as complex ecosystems on the seven continents and in our five oceans begin to unravel.

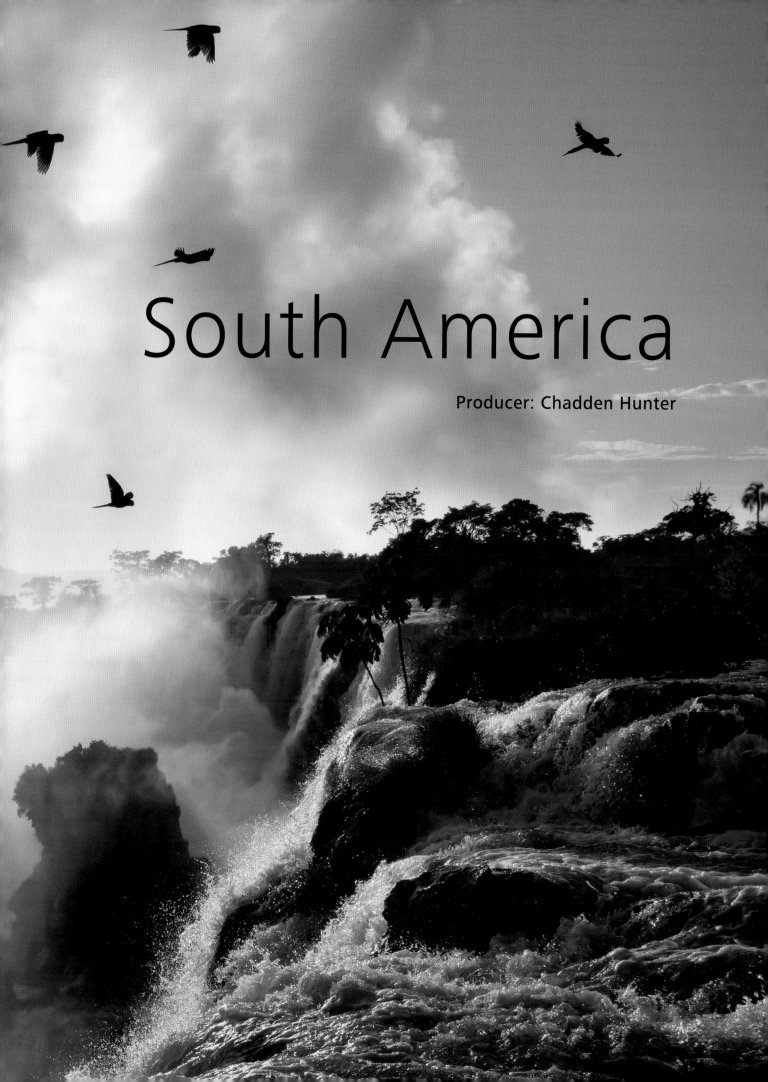

South America

Producer: Chadden Hunter

Introduction

South America, the fourth-largest continent, covers 12 per cent of the world's land surface, and is packed with geographical superlatives. The Amazon is the world's largest river, and the Andes the longest mountain range on land. Angel Falls is the highest uninterrupted waterfall, and Lake Titicaca the world's highest navigable lake. The Atacama Desert is the driest non-polar place on the planet, and Salar de Uyuni is the Earth's largest salt flat. In the south, the continent pushes farther south than any other, apart from Antarctica, so some of its upland areas have their own ice fields – the Northern and Southern Patagonian ice fields shared by Chile and Argentina.

South America is two-and-a-half times smaller than Asia, but it has almost all the same landscapes and biomes, which are home to more plant and animal species than anywhere else in the world. Today, much of South America's wildlife is unique to the continent, but there are distinct echoes of the other continents because the landmasses have been moving and a land bridge between South and North America has appeared and disappeared several times due to changes in global sea level. South America's marsupials are descendants of animals that were on the continent when it broke away from Africa, Australasia and Antarctica about 120 million years ago. Ancestors of New World monkeys probably island-hopped or rafted from the Old World across the Atlantic about 35 million years ago when South America and Africa were closer together. Predecessors of the continent's big cats, bears, guanaco, alpaca and vicuña likely moved south from North America two to three million years ago, when the Isthmus of Panama bridged North and South America. People probably came from Asia and migrated south through North America or arrived from Polynesia by sea. Nobody is sure which; maybe it was a bit of both.

A lost world

Sir Arthur Conan Doyle knew a thing or two about tepuis. His novel *The Lost World*, published in 1912, was set on the top of one. He wrote that it was a place where prehistoric animals survived and told how an expedition went to find them. While Sir Arthur's story was fantasy, the true story of tepuis illustrates how the real world can be even more engaging than a fictional one.

Tepuis (singular: *tepui*, meaning 'mountain' in the local Pemón language) are mainly to be found in Venezuela and western Guyana. Standing up to 3,000 metres tall, they are flat, tabletop mountains that stick out of the rainforest. They are made of Precambrian sandstone and are the remains of a large plateau that has eroded slowly down the eons to leave standing these huge towers with vertical cliffs. What makes tepuis biologically important, however, is that many of the plants and animals living on their flat tops, and in their extensive caves and wide sink holes, have been isolated from the rainforest below for thousands, if not millions, of years.

The caves themselves are unusual. They are formed in hard sandstone rather than limestone, so the erosion process is remarkably slow. The walls are lined with quartzite and the tunnels grow at about a metre every 100,000 years or maybe even more, nobody is really sure, so the vast caverns we see today must have been formed many hundreds of millions of years ago, and are possibly the oldest caves in the world.

Each tepui is also an ecological island, with unique flora and fauna, and the vertical cliffs, some 600 metres high, make it so difficult to explore here that many of its species are yet to be formally described by science. Many of the known species on Roraima tepui, for example, are unique to that mountain, such as the numerous species of carnivorous plants and an unusual member of the bush toad family. The pebble toad is not bothered by snakes up here, but there are large tarantula spiders roaming these tepuis, big enough to catch a small toad. If threatened, it pulls in its legs, tucks in its head, tenses its muscles and forms a bouncy ball so it can roll away from danger like a dislodged pebble, a form of defence rare in nature.

They all live in a very wet and windy environment, with extremes of temperature. On these tepuis, rain increases with altitude and, with so little soil to hold back the runoff, water levels in caves can rise 6 metres in a few hours. It spills into the surrounding rainforest, sometimes via spectacular waterfalls, like the Angel Falls or *Kerepakupai Vená*, meaning 'waterfall of the deepest place' – the world's highest uninterrupted waterfall.

The tepui can also be thought of as a microcosm of the entire continent. South America is species rich, so each must compete with many others for living space and food. Just like those on tepuis, the successful ones have developed specialist features to carve out a patch of their own.

Opposite
Tepuis are mesas composed of Precambrian sandstone. Auyá-tepui (top right) has Angel Falls, a 'plunge' waterfall with the highest uninterrupted drop in the world.

Behind the curtain

While Angel Falls is the highest waterfall, one of the most spectacular must be Iguaçu, on the Brazil–Argentina border, where up to 275 individual waterfalls combine along a 2.7-kilometre-long, horseshoe-shaped cliff to form the largest waterfall system in the world; and, while the falls themselves are stunning, some of the birds living here are equally impressive.

Swifts have established nesting colonies in an unusual place: directly behind the torrents of water pouring over the falls. It is certainly safe, probably the safest nest site in the world, but rather wet. Very few predators are likely to venture behind the curtain, but getting in and out is a challenge, as the birds must fly directly through a wall of water that drops 82 metres at an average rate of 1,756 cubic metres per second, and, during the wet season, it is even more difficult. It just shows to what lengths an animal might go to find a safe roosting or nest site.

The species is the great dusky swift, which is larger than most other swifts, and, unlike many of the others, it is often seen to perch and roost on

Above
Great dusky swifts roost and nest behind the cascades of water pouring over the Iguaçu Falls.

Overleaf
Iguaçu Falls is classified as a 'cataract' waterfall, in which a huge volume of fast-flowing water plunges over a cliff. It consists of many individual falls separated by islands.

exposed rock faces. Nests are disc-shaped, made of vegetation and mud, and are built on small ledges either behind the falls or on rocks between individual falls. Changes in water flow can sometimes mean the difference between breeding successfully and having the nest, eggs or fledglings washed away. The protection from predators, however, is worth the risk.

For those that nest behind the water, flying through it is just one problem. Returning parents must also be able to locate exactly the right spot in order to punch through the falls and land at the correct nest site, and their youngster's first flight is a bit of a nightmare. It has to fly at full tilt through the cascade, but they have never seen the other side. In what is tantamount to a leap of faith, the fledgling executes a small death-defying downward curve to minimise the impact of the falling water, and is free. Some, of course, do not make it, but the majority do and live on to hunt for insects in one of the last remaining patches of Atlantic Forest – the *Corredor Verde* – that surrounds the falls.

Above
Some swifts avoid the need to punch through the falling water by roosting on the rocks between falls.

Fish and fruit

The water at Bonito Pools is as clear as crystal, and some of the fish are as colourful as those on a coral reef. *Bonito* means 'beautiful', and it could not be more aptly named; a place of stunning caves, sink holes, and exceptionally clear rivers and pools. The rivers here flow through a karst region in southwest Brazil, and high levels of dissolved minerals, such as calcium and magnesium bicarbonate, help settle particulate matter, hence the clarity. Cameraman Bertie Gregory was taken aback.

'It's hard to describe how crystal clear the water is, but when you dive below it's surreal. It looks as if you're in a fish tank with a blue background, but pop up again and you're in a jungle river.'

The fish present include dorado, pacu and piranhas, but the production team were more interested in a large and colourful fish, known locally as the piraputanga. It can be up to 56 centimetres long and weigh 3.5 kilograms, and it reaches this size on a diet that includes a large quantity of fruit, which it acquires in two unusual ways.

Below
The water is so clear that a shoal of piraputanga in Bonito Pools looks as if it is in a sparklingly clean aquarium tank.

Shoals of these fish shadow capuchin monkeys that feed on the fruit in trees above the river. Wherever the monkeys go, the piraputanga follow. Capuchins are messy eaters, and quite a bit falls into the water. The fish are quick to grab every morsel. When the monkeys move on, the fish have a plan B. They leap up to a metre into the air and try to grab any low-hanging fruit for themselves. These specialised methods of feeding give them the edge over their many competitors.

It all sounds idyllic, but when Bertie flew his drone above the river, reality punched in.

'The protected tropical forest on each bank is little more than 100 metres wide on both banks. The rest had been cleared for agriculture, and we could see farmland way into the distance. To me the rivers were beautiful and full of fish but, chatting to the locals, I discovered that fish abundance was significantly down due to overfishing, and some rivers, which we didn't visit, were becoming cloudy due to agricultural runoff.'

It seems Bonito's beauty is becoming tarnished.

A lifetime in the wings

Along the east coast of Brazil lies more of the Atlantic Forest. With about 90 per cent deforested, it is a shadow of its former self, but there are still wildlife jewels to attract filmmakers. One little gem is the blue manakin, a tiny bird with a great performance. At breeding time, the male becomes an accomplished song and dance artiste, but he cannot stand out on his own. Instead, he gathers a troupe of male dancers – his support act – and they help him garner his special manakin, but they must all work extremely hard to impress her.

His body is an iridescent blue and he wears a bright red cap, so he stands out in the forest and, when the dance begins, he and his troupe put on quite a show. First, the group lines up along a branch. The leader is closest to the female, whose green plumage blends in with the trees. She looks on as, one by one, the males flutter flamboyantly into the air, facing the female, and then fly to the other end of the line. Director Maddie Close was in the audience.

'The males form a rapidly rotating conveyor belt, arriving in front of the female quicker than one a second. The line of four to five males may go through a dozen sets before the alpha male calls an end to the dance.'

The female seems mesmerised by the display, which can last for three or four minutes and, eventually, by some unknown signal, she indicates whether she is interested or not. Gradually, the support acts breaks up and flies off, leaving the leading male and female to consummate their union. It is a very specialised way to attract a mate and the rest of the hopefuls have to wait for the principal boy to pop his clogs in order to move up the theatrical hierarchy. About 90 per cent fail. They go through their entire life as permanent virgins – a lifetime in the wings.

Below and Opposite
A dance troupe of blue manakins is led by a dominant male. The troupe helps him to impress a female.

Life by the pool

Straddling the Equator in the northern half of the continent is the vast Amazon Basin with the world's largest tropical rainforest. Hiding there are some of the world's smallest frogs – the poison dart frogs. Their gaudy skin colour indicates that they are poisonous, and many are tiny, most species less than 1.5 centimetres long. They could sit on a thumbnail. There are larger ones and they are the frogs Amerindians use to tip their blow darts when hunting, giving the family its common name. There are also frogs in the family that have the warning colours, but they are just mimics and have little or no toxicity. Whether they do or not, they are all remarkably diligent parents.

The production team elected to follow a less poisonous species – the mimic poison dart frog. As its name suggests, it is not one of the highly poisonous species; even so, its skin is laced with potent pumiliotoxin B, which it probably gains from eating toxic insects.

It lives in the Peruvian Amazon, where it is active twice each day – early in the morning and again in the afternoon. For the rest of the time it retreats to a particular plant, but come the time to breed, which peaks in the rainy season, it shows how far poison dart frog parents will go to successfully rear a family. Few places are more crowded than a South American rainforest, and that means predators are rife and competition is fierce. The parents' showy colours might gain them a measure of protection, and they have a way to ensure at least some of their offspring have a fighting chance to make it to adulthood.

The male has a piercing call and inflates his black-spotted yellow throat pouch enticingly over the edge of a leaf to attract a female. She then follows the male to a suitable small pool within a plant, such as the water tank in the centre of a bromeliad, and she deposits a number of white eggs surrounded by clear jelly. A few days after the tadpoles hatch, the male carries them one at a time on his back, to be dropped into widely separated brood pools, which can be in different trees. The transfer avoids competition between tadpoles and is an insurance policy against predation. The tadpoles might feed on mosquito larvae, but the pools are so small that all the food is quickly consumed. This is where the parents step in again.

The male frog must remember all the locations of his offspring. He then leads the female to each pool in turn, and she lays infertile eggs that the tadpoles can eat. With this 'meals-on-legs' service, the tadpoles eventually metamorphose into miniature versions of their parents and they leave their pools to start their own life in the forest. Not all make it. Only two to four froglets reach maturity, out of the seven or so eggs the female deposits, so the survival rate for froglets is unusually good. It pays to look after the kids.

Life at a lick

Tambopata and Manu in the Peruvian Amazon are places to find large numbers of scarlet macaws – bright red, yellow and blue members of the parrot family. Macaws tend to be monogamous, pairs of birds are often seen flying over the forest canopy. They nest in tree hollows, where two to three eggs are laid. The chicks take a long time to fledge, maybe up to 90 days, and during that time the parents are challenged to give them a balanced diet. There are usually masses of fruit, nuts and seeds in the trees surrounding the nest, but there is a distinct shortage of sodium. The western part of the Amazon is sheltered from the Pacific Ocean by the Andes mountain chain, so there is little salt in the rain, and sodium is also leached from the soil. Youngsters need sodium for healthy development so, during the breeding season, the family's demand for essential minerals skyrockets. The answer is to visit clay licks.

The licks can be some distance from the nest, usually on the banks of the nearest river, and when the birds arrive they are extremely cautious. Many predators lurk in the forest, and, while the film crew were watching, an ocelot swam across to have a go at catching one. The birds wait in the trees next to the riverbank, waiting for a quorum of birds to gather before any drops down to mine the clay. Then, they all descend en masse, furiously ripping chunks of clay and flying off. It is, of course, what any predator hopes for. It is unlikely to show itself if there are just one or two birds at the lick. It stands a better chance of rushing one if there are many birds bunched together.

Above and Opposite
Red-and-green macaws and scarlet macaws, together with a sprinkling of blue-and-gold macaws, visit a clay lick in a riverbank in the Peruvian Amazon.

Almost homeless

The Colombian coastal rainforest is on the opposite side of the Andes to the Amazon. As in many parts of South America, it has suffered from serious deforestation. Colombia is said to be the second most biodiverse country in the world, but many of its species are endangered. The country has lost at least a third of its native forests, and much of the rest is threatened. Attempts to bolster the national economy are responsible: legal and illegal logging for timber, ranching, mining, hydroelectric power, roads, bridges, and even production of coca, the plant that provides the main ingredient of cocaine.

The coastal rainforests have been hit particularly hard. Rivers have been poisoned by mercury from goldmining activity and, because the trees have been removed and the soil is washed away, they are being choked by silt. Living in this ecologically hostile environment is one of the world's rarest primates – the cotton-top tamarin. The International Union for the Conservation of Nature (IUCN) indicates that it is Critically Endangered, with fewer than 6,000 individuals remaining in the wild. It once roamed the whole of these coastal forests, but now it is restricted to four small patches.

Cotton-top tamarins are among the smallest of primates, but they have a large brain, possess great agility, and communicate in whistles and chirps in such a sophisticated way that scientists consider it a language. They live in troops up to 13 strong, within which there is a distinct hierarchy: only the dominant male and female breed. The alpha female prevents other females from breeding by emitting pheromones, similar to the way queen honeybees suppress female worker bees in their colony. Like bees, the tamarin troop cooperates to bring up the young.

Food can be just about anything. These little characters are opportunistic omnivores, eating fruits, insects, small amphibians and reptiles, and even nectar from flowers and the natural gums that exude from tree trunks. As fruit is so abundant, it makes up about 40 per cent of their diet, making them important seed dispersers in the forest.

The tamarins are active during the day. Birds of prey and snakes are their main worries, along with small cats, such as ocelots and margays, and mustelids, such as the tayra. Troops tend to remain in the lower canopy and understorey, where they can hide amongst the foliage.

Loss of their lowland forest homeland, however, means that parent tamarins and their helpers must push their skills to the limit just to raise and protect the family. The youngsters themselves have an awful lot to learn about the troop within which they live and the forest they inhabit. They must copy cotton-top language and know their status in the troop, as well as learn how to clamber about in the trees, find out what is edible and where to find it, avoid being on a predator's menu, and – most important of all – how to stay clear of humans and the international trade in pet monkeys.

Opposite and Overleaf
Cotton-top tamarins are recognised instantly by the mop of long, white hair on their heads. The very fine facial hair makes it look as if they have naked, black-skinned faces.

Paddington and friends

The Andes mountain chain runs the length of South America, and at an altitude of between 800 and 3,500 metres in the tropical Andes lies a band of forest that is often enveloped in fog. This is the cloud forest. It has smaller and fewer trees than the tropical rainforest, and they are strewn with epiphytes – mosses, lichens, bryophytes, ferns, bromeliads and orchids – which seem to drip from the trees. Scientists are revealing that the cloud forests possibly have even greater botanical biodiversity than the rainforests.

One of their largest inhabitants is the Andean bear, formerly known as the spectacled bear and the wild inspiration for the late Michael Bond's Paddington Bear. It doesn't always have 'spectacles', so researchers changed the name. It is the only bear in South America, the last surviving species of short-faced bear and a living relative of the extinct giant short-faced bear of North America, the largest terrestrial mammalian carnivore that ever existed.

Below
The Andean bear is also known as the 'spectacled bear', because some individuals have white markings that resemble a pair of spectacles.

The smaller Andean bear is an inveterate tree climber, and even builds nest platforms in tree forks. Food is mainly vegetarian, often bromeliad hearts, although it will eat meat, and it has a particular fondness for the fruits of the pacche tree. It produces sweet, lipid-rich fruits resembling black olives that appear at some time between July and February, and whenever a tree sets fruit, bears arrive from all over the forest. It proved to be a boon for the filmmakers.

Opposite
When the olive-shaped fruits of the pacche tree come into season, bears arrive from all over the cloud forest.

While trekking through Ecuador's Maquipucuna cloud forest reserve in search of these apparently elusive bears, director Sarah Whalley and her film crew came upon a pacche tree and were suddenly brought to a halt.

'"Oso, oso!" cried our field assistant, and we realised there was an Andean bear ahead climbing the tree. In fact, we heard it before we saw it. These bears make such a loud sound with their claws, that it reverberates through the forest.'

And this noisy bear turned out to be the first of many that cameraman Bertie Gregory was able to follow and film.

'Andean bears are very rarely seen. Even the scientists who study them observe them only a few times a year, but Sarah had really hit the jackpot. On our first day we saw five, and on another day ten. Our consultant didn't believe us at first, but then we saw three large adults together in a single tree!'

The bears are more usually solitary, and although they do not have territories, they have large home ranges in which they keep themselves to themselves, so to have a bunch of them come down from higher altitudes

and turn up in a grove of these trees can spell trouble. These bears, though, *do* have a history. Some get along, others don't, the latter often coming to blows. Nevertheless, with such a powerful attraction as the pacche tree, it is not unusual to find several bears in a single tree; and they don't stick to the thick branches. Using their long, sharp front claws, they can haul themselves up and go to great lengths in order to gather the fruit, climbing to 20 or 30 metres above the ground. While Bertie was watching, he saw them moving towards the most precarious limbs.

'It was interesting how they appeared to understand the physics of weight and pivots. Once they'd finished eating the easy-to-reach fruit, they'd start edging out on the thinner branches. When they realised the branch could break under their own weight, they'd stop and chew off the end, but not completely. They'd snap it with a bit of a hinge still attached, and let it swing towards them. In this way, they could grab the fruit furthest from the trunk. All very clever.'

The bears might feed for about five minutes at one particular tree, and then move on to the next, often travelling across several valleys during the course of a day. They have well-worn trail networks between the trees, which are scent-marked by rubbing and scratching. They also gradually move to higher elevations as the fruits become progressively ripe through the season. When the pacche trees have finished fruiting, for example, the bears climb further up the slopes, where they switch to pasallo trees. They strip off the bark and feed on the sapwood inside, and maybe snack on land snails, insects, and wild bee honey… but not marmalade!

High plain

At elevations a little above the cloud forest – on average 3,750 metres above sea level – the Andes mountain chain is blessed with the spectacular landscapes of the Altiplano, the world's second-highest plateau. One of its most unusual features is located in southwest Bolivia. It is the Salar de Uyuni, the largest salt flat in the world. It is so flat, and has such clear skies, that the reflective white salt surface is used to calibrate the altimeters of Earth observation satellites. About 250 kilometres to the south, in northern Chile, is El Tatio, the largest geyser field in the Southern Hemisphere and the highest in the world. It reflects how young and seismically active these mountains are with some of the world's highest active volcanoes, part of the Pacific Ring of Fire. The summit of one – Chimborazo – is the farthest distance from the Earth's centre, sticking out even farther than Mt Everest's peak. This is because the Earth bulges at the Equator so, even though Chimborazo is a modest 6,263 metres above sea level, compared to Everest's 8,848 metres, it is actually closer to the sun.

Below and Opposite
The El Tatio geyser field is in the Andes Mountains of northern Chile. It's the largest of its kind in the Southern Hemisphere.

Overleaf
Salar de Uyuni in southwest Bolivia is the world's largest salt flat, the remains of a prehistoric lake that dried up.

Penguins in the desert

At lower elevations, the scenery can be very different, like the Atacama Desert, the world's driest hot desert. With much of the area sandwiched between the Andes and the Chilean Coastal Range, moisture does not reach the region from either the Pacific or Atlantic oceans. It is in a double rain shadow, and, having been extremely arid for at least three million years, it is possibly the oldest desert on Earth. Even today, some places have received no rain for hundreds of years, so they are totally barren. It is so dry that plants cannot grow; yet there are penguins wandering across the sand, as the desert west of the coastal range runs right down to the ocean and, along the shoreline, there is an extraordinary abundance of life.

The reason is that the Cold Humboldt Current, which flows northwards from Antarctica to the Galápagos Islands, kisses the Pacific coast of South America. Winds blowing parallel to the shore cause upwellings that draw up nutrients from the deep sea, the basis of a food chain that is bursting with life.

Punta San Juan in Peru is the section of coast that is closest to the deep sea, and it is claimed to be the most productive coastal marine site in the world. Peru's largest colony of Humboldt penguins is located here, along with very large populations of sea lions and fur seals. It is also one of the most important sites for the so-called guano birds – Guanay cormorants, Peruvian pelicans and Peruvian boobies – which once supplied huge quantities of nitrogen-rich guano to the fertiliser and explosives industries. Inca terns, Belcher's gulls,

Above
A 'wall' of South American fur seals and sea lions lies between the sea and the breeding sites of Humboldt penguins.

Overleaf
From the air, the demarcation between pinniped and penguin colonies at Peru's Punta San Juan can be clearly seen. Half of Peru's population of Humboldt penguins lives here.

snowy egrets, black oystercatchers and kelp gulls are amongst the supporting cast, but the stars of the show are the penguins. They are what the production team had come to film.

They nest and roost on the slopes behind the beach, hiding in gullies, crevices and burrows dug deep in the thick layers of guano lest birds of prey swoop in and grab them. However, the feeding is so good on this coast that it is worth the risk, and many pairs have two broods a year during a favourable breeding season that extends from early April to December.

In some years, though, very few or even no chicks are raised. These periods coincide with El Niño, when the currents reverse in the Pacific Ocean and warm water can replace the cold Humboldt Current. Fish head for deeper water until this weather phenomenon passes, but the penguins find it increasingly difficult to feed themselves let alone a family, so nests, eggs and chicks are abandoned.

In good years, fish are plentiful, but there is another impediment to a successful fishing trip. Between some of the penguin nest sites and the sea is a beach with hundreds of sea lions. They are packed so tightly together, there is barely a scrap of sand or rock visible. The birds have no choice but to clamber onto the seal lions' backs and, hopping from one to the next, make their way to the shoreline, while avoiding the snapping jaws. With fishing over, they have to do it all over again, except in the opposite direction.

Above
IUCN categorises Humboldt penguins as 'Vulnerable'. About 37,000 live in the wild, but each time an intense El Niño overwhelms the region, food is in short supply and chicks can be abandoned. Climate change could make the situation worse.

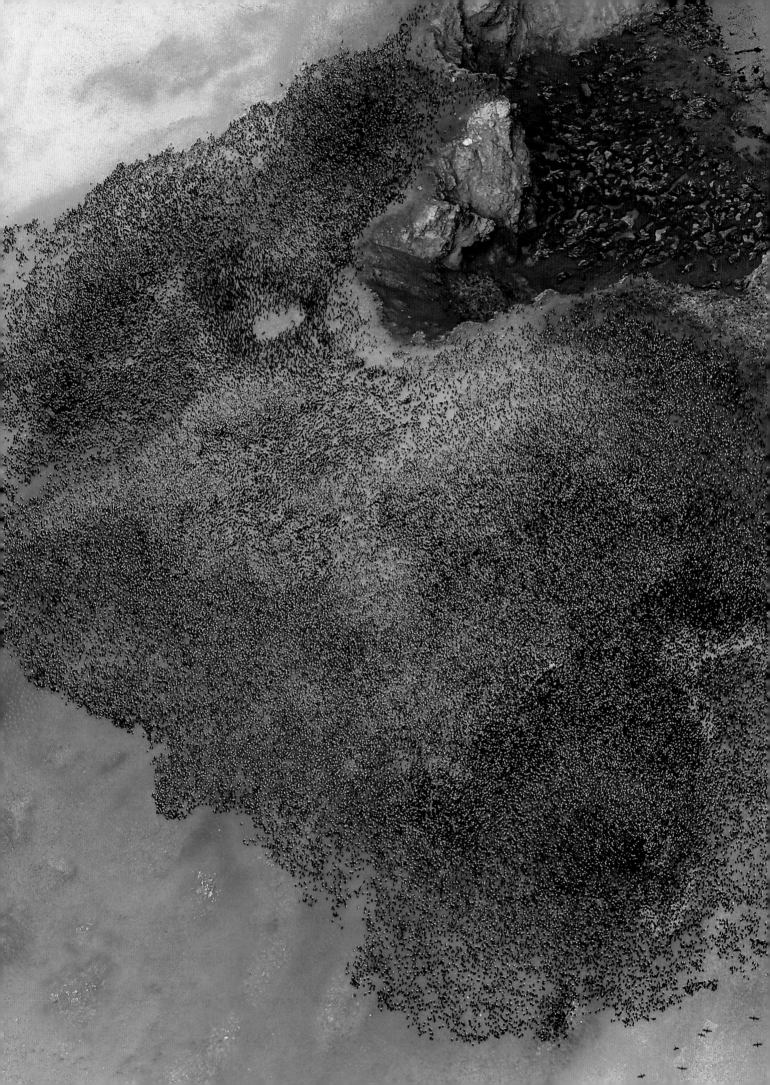

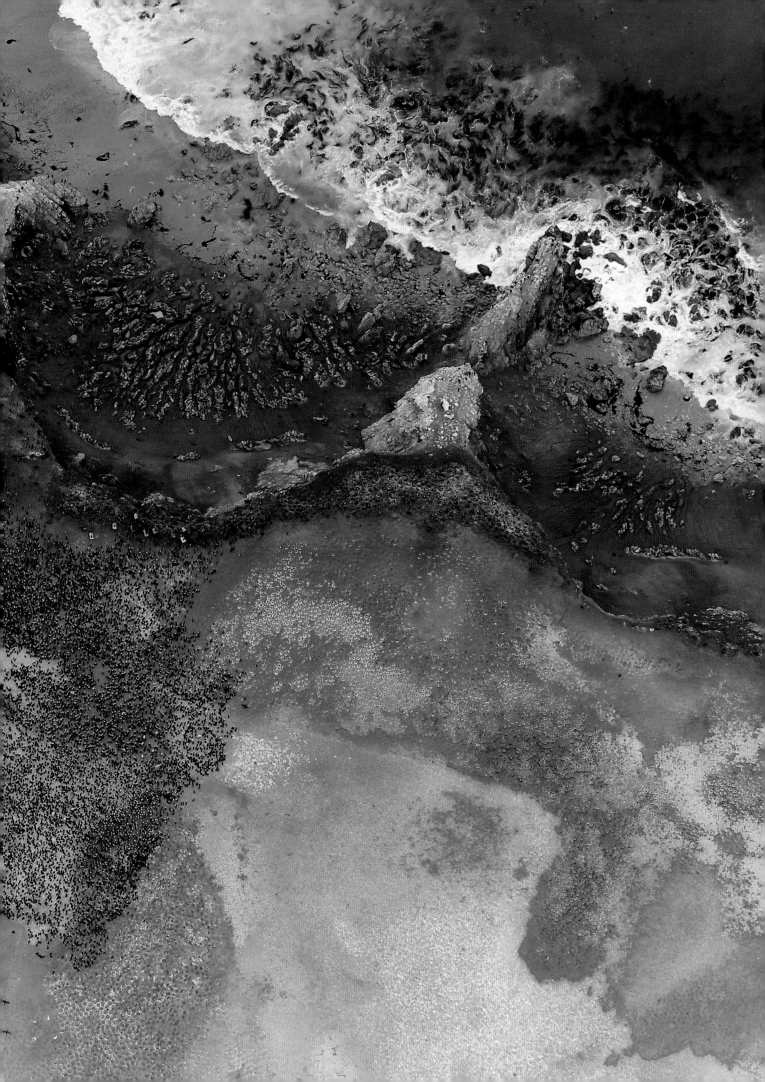

Behind the Scenes
The cat with many names

At the southern tip of South America, the landscape is nothing short of spectacular. The granite towers of the Parque Nacional Torres del Paine, for example, are part of one of the most stunning panoramas on the planet, and they are the backdrop to dramatic encounters between two of the continent's most charismatic animals – the puma and the guanaco.

The puma is a secretive but highly adaptive cat. It is found from Alaska in the northern hemisphere to the southern tip of Chile in the south, and in a range of habitats from high mountains to hot deserts and lowland tropical forest. It lives in 28 countries, where it has collected more than 80 names, the commonest being mountain lion, cougar and puma. Those in the Torres del Paine region are notable for being the biggest and most powerful individuals in all of the Americas.

The guanaco is one of the largest mammals in South America. It is basically a camel without a hump and the wild ancestor of the domestic llama. In spring,

Above
This is Sarmiento, probably the most filmed and photographed wild puma in the whole of the Americas!

its breeding season is unusually short, with rutting and calving at roughly the same time. Sarah Whalley was in Chile with the film crew observing the births.

'The baby guanacos are known affectionately as *chulengos*, and, although they wobble about at first, they can be up and running on their long, gangly legs five minutes after being born. For a dominant male, though, it's a busy time. He has to defend his harem from other males. If he spots a rival trying to steal one of his ladies, he lets out a high-pitched laugh as a warning, and his first line of defence is to chase the challenger away, but if that fails they come to blows.'

The brawls are intense. The two fighters rise up on their hind legs and slam into each other's chests, then they try to nip at their opponent's front legs in order to knock him off-balance… and even try to bite off his testicles. They are understandably distracted at this time of the year, and pumas take full advantage of their carelessness.

The puma is a supreme ambush predator. It can reach 50 mph in short sprints and has long and muscular hind legs, so it is a great leaper – 5.5 metres vertically and 12 metres horizontally. Powerful forequarters enable it to bring down prey much larger than itself. It leaps on the guanaco's back and tries to grab it around the neck and then simply hangs on tight, biting as high as it can to constrict the victim's windpipe and suffocate it. Guanaco, though, are powerful beasts and they put up a spirited fight. Many a life-and-death struggle ends with dinner escaping.

With such a formidable foe, mother pumas with growing kittens have their work cut out, and the film crew followed one particular mother for several weeks. Scientists studying her call her 'Sarmiento' and wildlife cameraman John Shier knows her well.

'The reason I became fixated with Patagonia was not simply to find pumas, but to locate pumas I could follow day after day, and this is *the* place for that. What makes it so special is a landscape with little tree cover and a core population of pumas that lack the shyness the species is famous for in the rest of its range.'

One reason for this, John discovered, is that the cats have become habituated to people, partly because a strong conservation movement has built in recent years. Ranchers around the national park have stopped killing the cats and are developing ecotourism. There is more money to be made from pumas than sheep, and so there are many more pumas around.

'I can say that without a doubt that some of the cats we filmed would have been dead had the ranchers not embraced the new ethic.'

And, one of those could well have been Sarmiento, but she is alive and well and, during the time that John and the rest of the film crew were following her, they found she had three large and very demanding kittens in tow, so she had to find sufficient food to satisfy all four of them. She was not helped when she took her kittens hunting and they jumped out of the bushes at the wrong moment and scared off their supper! There were other times, however, when she left her offspring in a safe place and went hunting alone. It tested the stamina of the film crew, for they could not simply drive up to where the cat

Above left
Guanaco mothers generally have one chulengo. If it's a male, it will be chased away from the herd around its first birthday.

Above
Sarmiento and her two kittens are heading out on a hunting lesson. The youngsters don't take part in the hunt at first, but watch and learn from a distance.

was hunting and film from the car or a hide. John had to follow Sarmiento on foot, whilst carrying a camera and tripod weighing more than 30 kilograms.

'We had the best guides and spotters – Diego Araya, Roberto Donoso and Marcial Urbina – who knew the cats well. Years of experience meant they knew what to expect, including hunting strategies. They kept us one step ahead of the game. We would drive to where they thought Sarmiento was, and then we hiked. I love it, but sometimes, I hate it; like when you've trekked for five miles up and over one mountain, only to see Sarmiento stalk a guanaco over the next steep hillside. Your legs and lungs burn, your back aches, and your body screams "stop", but you can't quit now that the cat is hunting. You pick up the pace and get up one more hill. With our excellent spotters and Sarmiento's tolerance of us, we spent almost every day with pumas.'

And, while John followed Sarmiento on the ground, Bertie flew his drone overhead.

'For tracking purposes it was useful to have the drone's eye view. I could guide John and our trackers, and also have an over-the-shoulder view of Sarmiento, often spotting the guanaco before she did. It was special seeing how her brain worked when stalking. She used the tiniest pieces of terrain to her advantage.'

Both the puma and the guanaco are considered 'Least Concern' in the IUCN Red List of Threatened Species. In the mountains, less competition with domestic stock and the new amnesty with ranchers means the guanaco and puma populations living in Torres de Paine are healthy. Guanacos are the most common large mammals in the park and the puma is their main predator, accounting for 74 per cent of guanaco deaths. The rest succumb to malnutrition and entanglement in fences, or they simply die of old age. The puma even influences their prey's behaviour.

In more open areas, where pumas are present, the guanacos bunch together in large groups, whereas in places with low shrub cover or low visibility they gather in smaller family groups. The more pumas are about, the more alert and flighty the guanacos are. They can run at about 35 mph, faster than any other Patagonian mammal… except the puma but, as Sarah found, the cats don't have everything their own way.

Above left
During the rut, male guanacos engage in often quite violent fights to determine who will take over a harem.

Above
Sarmiento uses whatever cover is available to get as close as she can to her target before making the final dash.

Overleaf
A filmmaker's dream: a guanaco herd socialising, dust bathing and generally chilling out in front of one of the most spectacular backdrops on the planet – the granite peaks of Torres del Paine, meaning 'towers of blue'.

'Following Sarmiento every day on foot, we felt her frustrations at failed hunting attempts, knowing she had hungry mouths to feed. She was a skilled hunter, but taking down an animal that weighs around 90 kilograms when you are only 40 kilograms yourself is difficult and dangerous. She came out of one failed hunt with a huge cut on her shoulder.'

This particular hunt came as a surprise to John. From his experience, pumas generally make a swift, clean kill, but this was something else.

'It was a bloody, all-out battle for survival. The guanaco stomped and kicked the hell out of Sarmiento, but what also shocked me was how relentless she was. It drove home the point of how difficult it is not only to be a puma, but also one with three nearly fully grown kittens to feed. If she doesn't lay it on the line like this, then the whole family dies. In spite of her tenacity, skill and aggression, she still failed, but she had to get up, dust herself down, and keep hunting.'

Sarmiento and her kittens, however, faced a lean period. For the film crew, it meant several weeks of failed hunts, but then, on the very last day of the shoot, their luck changed, and Sarmiento showed Bertie just how fast she could go.

'She hadn't eaten for a week, so we knew she had to be hunting soon – probably after we had left. We watched as she approached a couple of guanaco, but they got away. The light was fading fast, and Sarmiento had bedded down with her cubs. Our puma spotter Roberto looked at me and said, "I think that's us finished." I landed the drone and just took a few minutes to watch her and take it all in; that was when she suddenly stood up and stared into the distance. She began to run. I launched the drone and followed her progress, but I'd not seen her hunt like this. Usually it was a slow stalk and a quick sprint, but now she was running flat out for nearly a kilometre. Thanks to the drone, I could see the herd of guanacos she had spotted, and she could run undetected as they were over the brow of the hill and couldn't see her. But this was bad for me, as I'd lose contact with the drone. Gemma Templar, our drone spotter, and I had to hoof it to higher ground, while the drone hovered over Sarmiento. She was closing the gap to her prey, but we made it, just as Sarmiento popped over the ridge. I managed to keep her in frame as she chased down and hit a guanaco. Ten minutes later, we had recovered the drone, by which time it was too dark to film.

'It often seems to be a cliché for wildlife teams to say "and on the final day of waiting", but on this occasion, it couldn't have been more true.'

Australasia

Producer: Emma Napper

Introduction

Australia is the lowest, flattest and driest inhabited country within the continent of Australasia. About 70 per cent of the island is arid, semi-arid or true desert, mainly in the centre and the west, but there are tropical rainforests in the northeast, and snow-capped mountains in the southeast. Soils are nutrient poor and unproductive, yet the island continent is biodiversity rich, with many species found nowhere else in the world. It is also one of the oldest landmasses on Earth. Its rocks span over 3.8 billion years of Earth's history, and the world's most ancient material of terrestrial origin – a zircon crystal about 4.4 billion years old – was found in the Jack Hills of Western Australia.

Much of the island's soft, sedimentary rocks have been eroded away, and two of the world's largest inselbergs or island mountains – Mount Augustus and Uluru or Ayers Rock – are to be found in its dry outback. Both are sacred to the Aboriginal people who are local to each area. Their ancestors migrated here from Asia over 60,000 years ago, and maybe even 120,000 according to recent research, while the first Europeans arrived as recently as 1606, and promptly urbanised a fair bit of Australia's coast.

The continent's long-term isolation has resulted in an unusually large number of endemic species. Most of the native mammals are either pouched marsupials or egg-laying monotremes. At one time, Australia was warm and humid with widespread forests, but the climate became drier and most of the wildlife pushed to the edge of the island continent. The arid conditions favoured reptiles, with 755 known species, including five of the most venomous snakes in the world, which means Australasia has the most reptile species of any continent.

Food that drops out of the sky

During the dry season in Northern Australia, animals struggle with the intense heat and somehow survive on very little water. It can be six or seven months without rain and the temperature might be 46°C in the shade, but there is an escape from this dry and thirsty land if you happen to live near the Roper River. Fed by underground springs, it flows throughout the year, so it is a vital refuge for a large number of small mammals that normally shun the sun. They are little red flying foxes.

Bats, and flying foxes in particular, are among the few placental mammals living in Australia; most of the rest are marsupials or monotremes. The stretch of water between Southeast Asia and Australia was no barrier to them, and today the little red flying fox can be found in flocks and roosts containing upwards of 20,000 individuals, with occasional large groups up to a million strong. Unlike the larger flying foxes, which tend to stay close to the coast, this species moves further inland, and is nomadic. When the dry season comes along, they head for the Roper.

The bats swoop down to the river and skim across its surface, soaking the fur on their chests and then lapping up the water when back at their roost. It seems a simple task, but it is probably the most dangerous thing they will ever do. The river is home to Australian freshwater crocodiles and they have clocked that these little flying creatures would make a tasty snack. Cameraman John Shier was watching.

Above
A large group of little red flying foxes commutes between feeding sites, roosting camps, and the river. The bats feed on the nectar and pollen from *Eucalyptus* and *Corymbia* flowers.

'The thing that hits you is the angry heat. It's unlike anywhere else I've filmed, and it dawns on you why the bats take such a risk.'

Warmed by the summer sun, the crocodiles are alert and ready to catch the food that drops out of the sky. As soon as a bat is close to the surface, the crocodile surges out of the water and its jaws shut in the blink of an eye… but the bats are skilled flyers. They have to be. They must drink or they will dehydrate. It is a cruel whim of nature: while the dry season presents a hazard for bats, it offers crocodiles a heaven-sent opportunity. The crocodiles, though, are not especially good at catching the bats.

'On any given morning, maybe ten bats are taken,' John noticed, 'so you wonder why the crocodiles bother to congregate here. The key is that the crocodiles are adapted to survive on a sparse and intermittent diet. They excel at expending as little energy as possible, so if a croc eats a bat every few days, it's worth the time and effort to come here and hunt them.'

Below
The bats obtain water by skimming the river to soak it into their fur and then lapping it off later. Waiting for them are freshwater crocodiles or 'freshies', as they're known locally.

Overleaf
Little red bats live on the coasts of north and east Australia where they frequently decamp and roam widely to increase their chances of finding food and water.

The fisher king

Northern and eastern Australia is the haunt of the azure kingfisher. It is quite a striking bird, with a deep royal blue azure back, bright white spots to the throat and neck, a buff-orange chest, and red legs. Like many kingfishers, it dives for fish, aquatic invertebrates, and even small frogs in rivers and streams, but in locations with crocodiles it has to be a little more careful. In dark and murky waters, it is difficult to see what is what, let alone catch anything, but this kingfisher has adopted a friend to help it out.

Sinister bubbles appear on the water's surface. Is it a crocodile? Well, apparently not. The mysterious underwater bubble-blower is none other than the platypus, an animal that Victorian naturalists thought was a joke. It has a duck's bill, an otter's feet, and a beaver's tail, so it looks like several animals sewn together – a taxidermist's hoax – but it is actually a very sophisticated monotreme, a primitive mammal that lays eggs. For one thing, the male has venom glands linked to spurs on its back feet, making it one of the few venomous mammals. Both sexes have electroreceptors across the leading edge of their bills, which pick up the electrical activity in the muscles of their prey, making them accomplished hunters. The kingfisher has tapped into this. The bird follows the platypus wherever its goes, and is not averse to grabbing the invertebrates and other prey the platypus has disturbed on the riverbed, including the yabby, a freshwater crayfish.

Below
The azure kingfisher of north and east Australia and New Guinea feeds on small fish and freshwater 'yabbies', a type of crayfish.

Opposite
The duck-shaped bill of the platypus is lined with sensors that detect the electrical activity in the muscles of its aquatic prey.

High country surprises

Seeing a mob of kangaroos in a snowstorm is somewhat surreal. Normally, you'd expect to find them in the arid Outback, but in the Snowy Mountains of New South Wales kangaroos and wallabies are frequently spotted in the snow-covered landscape, albeit looking a little lost, their hopping somewhat curtailed. These mountains are the highest in Australia, with Mt Kosciuszko the tallest peak at 2,228 metres above sea level. They are part of the Australian Alps, where the island's lowest temperature was recorded. It was minus 23°C in Charlotte's Pass.

On the higher slopes of these mountains is another unexpected marsupial, the bare-nosed or common wombat. Like other wombats, it spends much of its life in a tunnel, with tunnels dug mainly on sloping ground so they drain well, but it eats more native grasses than the other two species. As a consequence, the common wombat is the only marsupial whose teeth grow constantly, like those of a rodent. In winter, it digs under the snow to locate grasses and forbs or searches for patches of grass and other plants without snow cover.

This wombat is more active at cooler times of the day, so it is more nocturnal during the summer, and is one of the few marsupials to be found above the snowline (see page 1). It is generally solitary with a home range that varies in size depending on food availability, although its range might overlap with those of other wombats. Interactions between neighbours tend to be aggressive. The wombat gives a low guttural growl as a warning but, when it is really riled, it emits a loud rasping, hissing sound accompanied by expelled air. Mothers and young exchange soft 'huh-huh' sounds, but how they ever reach the point of raising a family seems a bit of a lottery.

At breeding time, females are especially aggressive. In captivity, a female was seen to attack a male for about 30 minutes before mating, and in the wild, the female runs in wide circles while being pursued by the male. The male bites the female on the rump and rolls her over on her side, but after a few minutes she jumps up and the chase resumes. This can go on for a further half-hour, until the pair mates while lying on their sides.

Predators are mainly introduced dogs and foxes. If attacked, the wombat makes for its burrow, where it can use its rump to crush an assailant's head against the roof or walls of its tunnel. In the open, it can run away at speeds up to 25 mph, and it is tough enough to hold its own against a solitary dog, but a pack would be able to overwhelm it. They *can* be dangerous to humans. In 2010, a man was charged and knocked over by a wombat. It savaged his leg and left him with large scratches on his chest.

One last thing about wombats is the shape of their droppings – they're cubes! They use their poo to mark territories. The cube shape stops the droppings from rolling away, so the animal can stack them up. The excrement is moulded in the last 8 per cent of the intestines, where the elasticity of the walls means the droppings come out as two-centimetre cubes, unique in the natural world.

Opposite and Overleaf
While kangaroos in Australia's 'high country' often find themselves in a snowy landscape, those elsewhere rarely experience it. In recent years, however, unusual winter storms have seen kangaroos throughout New South Wales and even in subtropical Queensland having to survive sub-zero temperatures with falls of snow or sleet.

Dingo hunt

Dingoes are dogs native to Australia. Genetic studies indicate that they are related to the New Guinea singing dog, and either people brought them in boats to Australia 3,000–5,000 years ago or they walked over a land bridge at some time between 4,600 and 18,300 years ago when sea levels were lower than today. They were never really domesticated, so they are not considered feral dogs, and since they have been in Australia they have evolved into the wild dingoes we see today.

They either live and hunt alone or in small packs, and may roam great distances in search of food. They will tackle an adult kangaroo, but a baby or juvenile is easier to catch, so, during filming, when a small group of dingoes appeared over the hill, all the kangaroos panicked, no matter their size. Joeys leaped into their mother's pouches, and they all set off at high speed.

Kangaroos can bound along at about 25 mph for long distances, with bursts up to 40 mph, while the dingo can make a top speed on average of 30 mph; so the predator can just about keep up with its prey, although adult kangaroos still have the upper hand. On this hunt, the pack focused on juveniles, as they were too big to hide in their mother's pouch. The youngsters had to outrun the dogs.

The chase over the uneven ground was dangerously fast. A juvenile was cornered but it escaped by leaping right over its assailant, twisting in mid air. The other dogs cut corners and tried to intercept the fleeing kangaroos, but they caught nothing, returning to their den site with their tails between their legs.

Below and Opposite
The dingo mother spots, stalks and then pursues a young kangaroo. The youngster should be able to outpace the dog, but if it trips and falls, it pays for its mistake with its life.

At the den the cubs waited for food. Usually but not always, the alpha female – in this case a white dingo – bears cubs, although the rest of the pack will help to feed them. This time they had come home empty-handed, but their mother was not content with leaving it at that. There were young kangaroos out there; it was just finding a way to trap one. She headed out again, but this time alone.

She found the mob not far from where she had chased them. They were alert, and agitated. Using what little cover there was, she began to stalk, slowly easing her way towards a group of juveniles. Suddenly, she burst from cover, targeting one particular animal. It sped away, faster than the dog, but it stumbled. Quickly regaining its feet it raced off a second time, but stumbled again. There was no third time. The cubs ate that day. Their future, however, depends on where they end up living.

Australia seems to have a love–hate relationship with the dingo. It cannot decide whether it is a pest or needs protection. In one state it is protected and in the next it is shot. It has been little studied, as it is generally shy of people. Its genetics have been well worked out and argued over, but its behaviour and ecology in the wild are still relatively unknown. Nevertheless, executive producer Jonny Keeling, who was in the Outback with the film crew, found that science is beginning to be more sympathetic towards the species.

'The interesting thing about dingoes to me is that scientists now believe, ironically, that where they occur, they actually increase biodiversity, and they do this by keeping down feral cats, foxes and other introduced species.'

Dingoes, it seems, are getting a second chance.

Left
A pair of male kangaroos 'box', a ritualised form of fighting that often occurs when young males are in the vicinity of females in oestrus. Older males with females tend not to get involved.

Below
With Australia's high country as a backdrop, the broad sweep of grasses and gum trees becomes a vast arena in which a daily drama, featuring dingoes and kangaroos as predator and prey, is played out.

Peekaboo spider

Living in roughly the same area as the dingoes is another female hunter. She is a 5-millimetre-long jumping spider and a ferocious mini-hunter, so when the time comes to look for a mate, males are especially wary. They play a game of peekaboo, and they do it in an unusual way.

The male has heart-shaped paddles on the third of his four pairs of legs, and he uses them to signal to the female. He seems rather reticent to face her head-on, so he hides underneath a leaf and just waves these soft, fine paddle-shaped bristles in her direction. If the female has already mated or is unimpressed, she may attack him. Despite the danger, he might try again and again, flashing his strange appendage for hours before the female moves off or the male gives up, but an unmated female is usually transfixed when he flashes his paddles at her. He then sneaks up on her side of the leaf and mates with her, before beating a hasty retreat.

The scientist who revealed this remarkable behaviour is Jürgen Otto, who is a mite expert in his day job for the Australian Department of Agriculture, but a spider fanatic in his spare time. His discovery was a stroke of luck. He spotted the spider on his tent bag, but was not sure if it got there from his garden or from where he recently camped. He went back to look and found more, and then watched how the male uses his 'paddles' and the female responds. He believes the males can work out through leaf vibrations whether it is worth trying to mate. Their whole performance is to find a female that will not attack them.

Above
While on the underside of the leaf, the male jumping spider shows his paddle-shaped leg in order to attract the attention of a female on the topside.

Opposite
The male is very wary of the female. He could lose his life. He picks up her mood and intentions via vibrations in the leaf. He must find a female that is unlikely to attack him.

Courtship train

Echidna males have an equally frustrating time. They are not attacked by the females, but they are led a merry old dance before they can mate with her.

The echidna, like the platypus, is a monotreme. The female lays eggs like a reptile, but produces milk like other mammals. The four living species fill a similar ecological niche as hedgehogs in Europe, and they even look like them with a spiny back, a long snout and short legs. They eat mainly ants and termites or worms and insect larvae, depending on the species, hence their alternative English common name: spiny anteaters.

Short-beaked echidnas, which live in Australia and Tasmania, do not tolerate extremes of temperature, so they will retire to a cave or into crevices

amongst rocks to escape the worst of the weather. In winter they hibernate, when their body temperature can drop to the same temperature as the soil, from a normal active temperature of 30–32°C down to 4.7°C, according to research at the University of Tasmania.

For most of the year these animals are solitary but, in spring, when a male spots a female, he gives chase. She, meanwhile, has no intention of hanging around. She runs with a characteristic waddling gait. Her body rolls and yaws, so she achieves a relatively low top speed of 1.4 mph. Even so, the male in hot pursuit struggles to keep up with her, and he is not alone. As they trundle through the scrub other males join the party, each one trying to be closest to the female. Rough and tumble fights can break out, but the males do not want to be left behind. They cannot stop even to eat or drink. They can only take a break if the female comes to a halt, because she is only receptive for one day in the year and each male wants to be in pole position when the opportunity arises. If not, it could be a long wait until next time!

Wild budgies

Vast swathes of Australia are dry. About 20 per cent of the land is true desert, but that figure rises to 70 per cent if areas receiving less than 500 millimetres of rain a year are included. There *is* water – nowhere has less than 100 millimetres a year, compared to other deserts that can sometimes have none – but the wildlife must discover where it is in sufficient quantities to quench their thirst. One familiar creature is very good at doing just that.

The budgerigar is a long-tailed parakeet native to Australia. The wild birds have yellow plumage on their head, yellow breast feathers, dark blue tail feathers and black markings. They are quite small and seemingly delicate birds but, as one of the production team said, 'They're as hard as nails!'

The wild budgie lives in open habitats, such as scrublands, open woodlands and grasslands, where it is nomadic, constantly searching for food – mainly seeds – and water. Travelling flocks tend to be small but, when open water is found, birds arrive from all around, so these temporary mega-flocks can number more than a million birds... Even so, they are hellishly difficult to find, as director Lucy Wells discovered.

'You'd think finding a flock of hundreds of thousands of bright green birds in a big open landscape would be easy, but how wrong I was! I've chased budgies the length and breadth of Australia for 18 months. The problem is the big flocks break up into smaller groups, and they're always on the move, spending just a few days in one place before seeking sustenance elsewhere. Finally, we had a successful shoot, but I had organised the whole thing from Bristol with an Australian film crew. After all that searching, I never got to see these huge flocks of birds in the wild!'

The flocks themselves have rules, at least for part of the time. When the budgerigars are in the air, individuals avoid colliding by always turning to the right, but in the water it is every bird for itself. The birds seem to throw themselves in, and there are so many that they clamber on each other's heads and grapple with their claws. They have to watch out, though. It is not only budgies that are attracted to the waterhole: predators are there too. Goshawks and black kites swoop in and try to snatch birds from the wheeling, bathing flocks. Sheer numbers confuse the hawks, but one or two budgies still make the ultimate sacrifice.

Below
Large flocks of wild budgerigars attract attention. One of their main predators is the Australian goshawk. It's hunting a flock in flight at Shire of Murchison, Western Australia.

Opposite
Wild budgerigars are green and yellow with black markings. The other colours, such as blue, white and grey, are only seen in captive birds. Their feathers fluoresce in ultraviolet light, which might have something to do with courtship and the selection of a mate.

Overleaf
The greatest numbers of wild budgerigars are usually seen when several smaller nomadic flocks converge on a watering hole.

Little devils

Islands, whether they are surrounded by water or are isolated populations of plants and animals, can be sanctuaries or death traps for wildlife. When rats, red foxes or feral cats get a toehold, the native wildlife inevitably declines, but keep these characters out and it has a fighting chance.

This has been the strategy in the fight to save the Tasmanian devil, a marsupial predator that has been hit by a particularly nasty facial tumour disease. It threatens the survival of the species, but on isolated islands off the coast of Tasmania, disease-free devils have been released. These 'insurance populations' have been isolated from people, the thinking being that if the Tasmanian devil becomes extinct on the mainland, the disease will become extinct too, and the island devils would be in the vanguard of animals reintroduced to their former range. Lucy Wells went to see them.

'I was surprised how small they are: no bigger than a small domestic cat, and they are much more gentle and shy than they are usually portrayed.'

Lucy was there to film a new family of 'Tassies', as they are known locally. The youngsters, she learned, are called 'imps'.

'They were about the size of a kitten, and as bouncy as one too. Each had its own personality: one was really bold and was always the first out to play,

Opposite
Baby Tasmanian devils have to fight to become juveniles from the moment of birth. Their mother only has four nipples, so competition is fierce and some newborns will not survive if there are more than four in a litter.

Below
These young devils will grow up to have a large head and neck, which enables them to have the strongest bite for their size of any mammal land predator.

another was really chubby and very shy and, like most young mammals, they often seemed not to have a clue. They were scared of everything, even a rock, and reacted to any strange noise, including local birds calling. They were unsteady on their feet and often had hiccups!'

The family had four imps, and their mother seemed reluctant to leave them in order to go foraging. Was she aware that some were confident, while others needed some mollycoddling? The different personalities that Lucy had noticed have been the subject of scientific studies by researchers at the universities of California, Davis and Tasmania. It seems the Tassies' boldness or shyness, rather than their physical strength, determines how they will live their lives and for how long. Survivors are more than three times bolder than those that succumb to the challenges of everyday life, and that has a direct bearing on which animals will thrive or not after translocation. By the time the film crew were ready to leave, however, the four imps had grown significantly.

'They were so much braver,' Lucy observed. 'They no long played nervously close to the den. At sunset, they were off exploring the forest.'

Tassies are nocturnal animals, and they feed on anything that is seasonally abundant. Meat is generally in the form of carrion, but they do hunt and will take down a small wallaby, although young wombats are preferred because they have more fat on them. Home is one of three or four dens, ironically tunnels dug by wombats. Today, the species is considered endangered, but those populations in protected areas are beginning to recover.

So far, the island devils are doing just fine.

Right
The Tasmanian devil is a carnivorous marsupial related to quolls. It was once found all over Australia but it is thought that it was wiped out by dingoes and human persecution. It now resides only in Tasmania and a few offshore islands.

Swim-through cleaning services

Ningaloo Reef on the Coral Coast of Western Australia is one of the world's largest near-shore fringing reefs. It is famous internationally because whale sharks visit each year, and visitors come to dive with them, but this 260-kilometre-long reef has much, much more. It is rich in coral, fish, crustacean and mollusc species and its beaches are breeding sites for loggerhead, hawksbill and green sea turtles. The short-nosed sea snake, which was thought to be extinct, was rediscovered here in 2015, and sponges found in a deeper part of the reef are entirely new to science. The reef is also on the migration route for humpback whales, dugongs, dolphins and manta rays. All in all, it gives the Great Barrier Reef, on the east coast, a run for its money.

One of its residents is the 1.8-metre-long grey reef shark, a species that patrols the edge of the reef, where it drops off into deep water. At Ningaloo, they often visit a special section of the reef and stop off for a bit of TLC. Most cleaning stations require the customer to drop down and be immobile while the cleaner fish or shrimps do their work, but grey reef sharks cannot do that.

The sharks are obligate ram breathers, so they must maintain a forward swimming motion for water to pass continually over their gills in order to obtain enough oxygen. If they stop, they sink and drown; so, the grey reef shark cleaning station must be at the edge of exposed oceanic reefs, like the reef at Ningaloo. Here, a significant current enables them to hover, yet maintains the flow of oxygen-rich water over their gills.

At any one time, as many as 18 male and female sharks might form an orderly queue for the wash and brush-up service. They circle until it is their time to drop in, and as many as 12 may be cleaned at the same time at a 'bommie' or submerged outcrop of coral reef about 6 metres wide. The sharks initiate the process. They face into the current and hang almost vertically in the water with their heads and open mouths uppermost. It is a signal for blue-streak cleaner wrasses to come out of their hiding places amongst the *Turbinaria* corals and divest the sharks of external parasites and dead skin. The fish even enter the sharks' mouths and clean between those razor-sharp teeth… yet they exit intact and alive!

Above
The grey reef shark must keep swimming in order to breathe, so it can only stop at a cleaning station if there's a strong current that sweeps water over its gills.

Opposite
The bluestreak cleaner wrasse has a distinct and easily recognisable colour pattern of black stripe along its side with a white belly and back, and iridescent blue towards the tail. It's like a red-and-white barber's pole only for fish.

West coast bait ball

Spruced up and ready for action, the grey reef sharks use their keen senses to locate food, and a commotion in the water is a sure way to attract their attention. From time to time on Australia's west coast, the entire ocean seems in turmoil. The cause is a humungous bait ball of small fish. There are so many packed so tightly together that the sea turns black. It is shark paradise.

The sharks push the fish towards the shore, although the manoeuvre seems not to be coordinated, but when the shoal reaches shallow water, the fish are trapped. It is the signal for the sharks to attack.

The fish squirm this way and that, forming great seething spirals, each fish trying to be in the centre of the ball and desperate to avoid the snapping jaws. The sharks pile in, at first just fifty or so, but the word quickly gets out. The erratic movements of the fish produce low-frequency sounds that carry for some distance underwater, and the sharks pick up every vibration, so they begin to arrive from all around. In no time at all, there are more than a thousand sharks, one of the greatest concentrations of sharks ever seen on this coast.

Below
When viewed from above, a bait ball of fish is packed so tightly that it looks like an oil slick.

Opposite
This is what Western Australians call a 'chop up'. Huge numbers of sharks have arrived to feed on the dense shoals of fish.

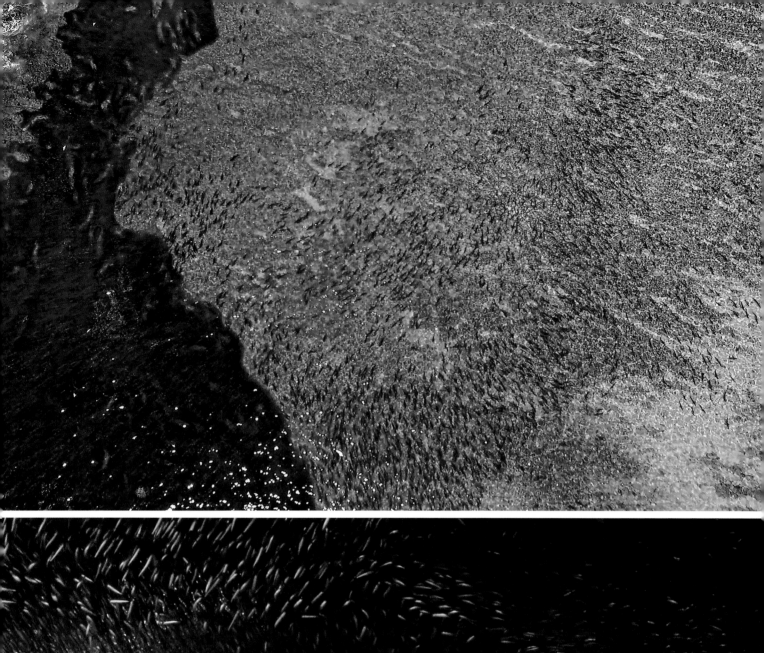

Behind the Scenes
Queensland's velociraptors

Daintree Forest is very old. It is located in Queensland, northeast Australia, where it is part of the largest rainforest in Australia – the Wet Tropics Rainforest – and the oldest continually surviving tropical rainforest on the planet. At one time, Australia was more humid than it is today and tropical rainforest dominated, but, as the climate changed and it became more arid, few rainforests survived. The climate and topography of the Daintree region, however, was just right, and this rainforest became a refuge for species with an ancient past, such as the idiot fruit tree – known as Australia's 'green dinosaur' – whose ancestors were here at least 120 million years ago, the date of the oldest known fossil.

The forest itself grows right down to the sea, and, walking along the beach, you might be taken aback to see fresh dinosaur footprints in the sand; after all, they are supposed to have died out 66 million years ago… or did they? Biologists consider birds to be living dinosaurs, and one of the most dinosaur-like is Australia's cassowary, the animal responsible for those enigmatic footprints.

The cassowary is a large and powerful flightless bird, females bigger than males. They can stand up to 2 metres tall. Both sexes have a casque covered with keratin on top of the head. Its function is not totally clear, although several uses have been suggested: it acts like a radiator to help regulate temperature; it amplifies the bird's low-pitched calls; or it is a visual display during courtship. Nobody really knows, as the three species of cassowary have been little studied.

One reason is that they are unbelievably shy, and run – at speeds up to 30 mph – deep into the forest at the first sign of people. What we do know is that a female has a territory that encompasses those of several males, for she mates with them all, but leaves their fathers holding the babies. Male cassowaries look after the eggs, which are laid in a pile of leaf litter, and when the chicks hatch out, fathers are generally good parents. The production team, however, found that the cassowary's shyness made it a difficult subject to follow in the wild.

Most wildlife films of cassowaries have been of animals in captivity, but the British producers teamed up with the only Australian camera crew with permission to enter a restricted part of the forest where these rare birds are living. They opted to follow a male with chicks, and there was one male, known to the team, which would tolerate people… but first they had to find him.

The first stage was to put out camera traps in parts of the Daintree Forest where the cassowary father might be roaming, but the team were unprepared for what they discovered. The biggest surprise was the feral pigs: they were enormous! They looked about the size of small ponies, but fatter. They are just one of the animals that should not be there. They churn up the forest floor, destroy native trees, and, when torrential rain falls on the forest, their

Opposite
The southern cassowary has a large skin-covered keratin casque on the top of its head. Its function has been long debated, but two recent proposals suggest that either it acts as a thermal radiator to lose and gain heat, or it helps to amplify the bird's very low-pitched calls. These sounds travel great distances through the rainforest, so these solitary birds can find each other at breeding time.

foraging activity causes flash floods, landslides and mudslides. With the soil removed, trees struggle to grow. A second surprise was that a camera trap had photographed a dingo, a species that had not been known in this area before.

There was still no sign of the father bird. More searches and camera traps followed, and eventually he was located. Directing the sequence was Lucy Wells.

'My first encounter with Crinkle Cut – named for his crinkly casque – was when he caught me unawares and gave me a beady look. He was letting me hang out in *his* forest, and over three field trips we got to know him very well.'

It was November when the film crew went into action, but the crew had a major disappointment: the brood of that year was lost.

'He was a very caring and extremely gentle father, and it was lovely to see his chicks grow. When they died, we were devastated. It's always hard on a shoot when tough things like this happen, but this was harder than most, as we had spent so much time with him.'

There would be no more youngsters until July the following year, but that season proved a disaster too. On the third attempt, the father was more successful and reared two chicks.

Despite the bird being used to them, the production team had to be careful not to upset him. Cassowaries sometimes attack people, especially when their offspring are compromised or when feeding sites are jealously guarded, and their assaults can be vicious. The inner toe on the bird's foot has a large dagger-like claw, up to 10 centimetres long, and this can do much damage. One well-placed kick from its dinosaur-like feet and its claw could disembowel a person unwise enough to get too close, although there are no recorded attacks where this has occurred and there are only a couple of documented

Above left
Finding a cassowary in the depths of the Daintree rainforest is not easy, but after spotting Crinkle Cut on camera traps the film crew was able to begin filming.

Above
The male cassowary is left holding the baby. It is up to him, rather than the female, to ensure their chicks come to no harm.

fatalities. In 1926, two teenage brothers came across a cassowary on their property and proceeded to attack it with clubs. One brother was kicked but ran away, and the other tripped and fell to the ground, whereupon the bird kicked at his neck. The claw sliced through the boy's jugular and he bled to death. The only other known death was in April 2019, when a captive cassowary killed its owner after he fell down in the backyard of his home in Florida.

The southern cassowary is relatively rare. It is endangered in Queensland, although, because it is mainly solitary and so difficult to count, nobody knows how many there are – perhaps, 1,500–2,000, but these are just educated guesses. The primary cause of its decline here is habitat destruction and fragmentation, but an enormous number are killed in collisions with motor vehicles and attacks by domestic dogs. Feral pigs destroy nests and churn up the forest, and tropical storms flatten the bird's habitat, but this bird is an important rainforest resident.

Adults eat mainly fruit, which they swallow whole, and one fruit – the cassowary plum – is named after the bird. In fact, adults feed on the fruits from many rainforest trees and so they are important as seed dispersers. The seeds are egested in the birds' muddy purple droppings, which are usually studded with all kinds of seeds and fruit stones. They can be deposited, along with a dollop of natural fertiliser, up to a kilometre from the parent tree. One tree – the rare *Ryparosa kurrangii* from the coastal rainforest – has a more successful rate of germination if its seeds have passed through a cassowary's gut, as do many others. It makes the cassowary a keystone species. Large swathes of Daintree rainforest trees depend on it, along with all the other plants and animals that rely on the trees.

Above left
The chick has to learn from its father what is edible and what is not.

Above
The film crew must be careful not to alarm the cassowary. Its powerful feet and claws have been known to injure and even kill a person.

North America

Producer: Chadden Hunter

Introduction

Occupying 16.5 per cent of the world's landmass, North America is the third-largest continent. Like Asia, it extends from the Arctic to the tropics, but has the added excitement of extreme weather. It's all down to its geography.

Unlike Asia, North America's principal mountain chains run generally north–south. In the west are the Rockies, Sierra Nevada and Cascades, with the Sierra Madres extending into Central America, and in the east the Appalachians. In between are the vast grasslands of the Great Plains, and a complex matrix of wildlife habitats ranging from steppe in the north, hot deserts in the southwest to vast swamps in the south and southeast. This means that, with nothing to impede it, cold polar air penetrates deep into the south, where it collides with warm air from the tropics and creates meteorological mayhem.

Northern Texas, Oklahoma, Kansas and Nebraska are the key states in what has been described as 'Tornado Alley'. The USA is plagued by up to 1,717 major tornadoes a year, more than anywhere else in the world, with some tornadoes reaching the southern Canada prairies. Wind speeds in the vortex can be up to 318 mph, as they were in an Oklahoma City tornado on 3 May 1999, the highest ever recorded in a tornado; and, if that's not enough, the southeast of the USA and its Gulf coast are hit regularly by another of Mother Nature's body slammers. They are in the direct path of extremely powerful storms forming in the tropical Atlantic. An average of ten named tropical storms, of which six become hurricanes with sustained winds up to 190 mph, cause death and destruction in North America and the Caribbean each year. The frequency of these two extreme weather systems makes this continent unique. It also means that some plants and animals – like their human counterparts in the eighteenth and nineteenth centuries – have to be pioneers, pushing themselves further than others of their own kind in order to find an unoccupied place that they can call 'home'.

Snow cat

The Rockies not only channel the cold Arctic air to the south; they can also protect against it. They shield the slopes on the Pacific side of the mountain range from the icy blast that barrels down the middle of the continent. It means that the Canada lynx, as long as it stays on the western side of the mountains, can venture further north into the Arctic than any other cat on the planet. Equally at home here is its main prey: the snowshoe hare. In winter, their habitat is dominated by heavy snow. Moist air from the ocean dumps

huge quantities of snow on the mountains, but the lynx and the hare are adapted to live in such conditions. They are both North American 'pioneers'. Producer Chadden Hunter and his film crew, on the other hand, were not!

'Winter in the Yukon delivered the most brutal conditions we encountered while in North America. It wasn't just that it was minus 30°C and often a howling blizzard. The snow was so soft underfoot – nearly two metres of powder covering a tangled mass of logs and fallen branches – that it meant

Above
With its thick fur coat, the Canadian lynx is at home in the snowy Rockies.

every step with a snowshoe was a test of strength and nerves. Most of the crew ended up with twisted knees and bruised sides, and we spent days and days following lynx footsteps in the snow, without us ever seeing them. We knew the cats had been watching us, but when we finally did catch a glimpse, we realised that we had habituated a wild lynx to our presence. To see those piercing yellow eyes staring back at us from only a few metres away was simply breathtaking. One of North America's most elusive creatures had let us into *its* world.'

The cat, twice the size of a domestic cat, has pancake-shaped paws that can spread as wide as ten centimetres. This enables it to pad around on top of the snow, and it has fur on its soles that help it grip. Similarly, the hare has snowshoe-shaped feet, hence its name. They are well matched, but the hare has snowy-white fur that blends perfectly with its background. As long as it remains still, it is safe, but if a predator approaches too close, the hare suddenly bursts from its hiding place and moves rapidly across the snow. With such a speedy animal, the lynx must use stealth and a surprise attack to have any hope of getting within striking distance. It does this by hunting mainly at dusk and during the night, when it might stake out a regular hare pathway or

places where hares gather. It tracks the hares using vision and a keen sense of hearing. It then pounces on its chosen victim, despatching it with a bite to the head, throat or the nape of the neck.

The lynx needs to catch a hare every day or so to meet its energy requirements and to stay healthy; but it is only successful in catching prey during a third of hunts. This rises to more than half if two lynx hunt together. More usually, though, they live and hunt alone.

Most years, they must travel long distances to find sufficient prey, up to 5.5 kilometres a day, but the hare population increases dramatically every 8 to 11 years. With food so readily available in boom times, the lynx has no need to move far. In bad years, however, the hare population can drop from 2,300 per square kilometre to as few as 12, so the lynx must seek other animals, such as ptarmigan, squirrels, voles, and young Dall's sheep and reindeer, and they are not averse to scavenging roadkill. It might also travel up to 250 kilometres away from its core area. Yet, despite all these challenges, this secretive cat is so beautifully attuned to its environment that it not only survives, but also thrives in the long cold winters and brief warm summers of the Yukon.

Above
The lynx's cryptic colouration enables it to blend in with the tangle of snow-covered undergrowth.

Beachcombing bear cubs

On the coast, to the west of the Rockies, the close proximity of the Pacific Ocean results in mild summers and relatively warm winters with high rainfall. It supports the largest expanse of temperate rainforest in the world. Stretching from southern Alaska to northern California, the forests are home to the North American black bear; and on Canada's Vancouver Island there lives a subspecies that is slightly larger and blacker than its mainland relatives. The bears spend much of their time in the island's lowland forests, but in spring, especially at low spring tides, they take advantage of what the ocean has to offer.

Here, the tide goes out to a great distance because of the gently shelving nature of the beaches, so bears seek out new feeding opportunities along the shore. Mother bears are often the pioneers. In spring, they emerge from their winter dens close to the shoreline with cubs that were born in the depths of winter, and they head down to the shore. It's beach school for bear cubs, and watching their progress was director Maddie Close.

'It was 4 o'clock in the morning when we first entered the inlet, and there they were – a large black dot, followed by two smaller dots lumbering slowly

along the shore in beautiful golden light. As we approached, the mother looked up and sniffed the air, but the cubs didn't seem to know that we were there at all.'

Joining Maddie was cameraman Bertie Gregory, and they followed this same mother and cubs for a month, during which time they figured out her daily routine.

'We filmed them each day at low tide, when they came out of the forest. They had been eating berries, but on the beach they switched foods. The retreating tide revealed a vast boulder field and under the stones was a seafood feast.'

The cubs, however, had to learn how to get at it, so their first lessons were to know where to look and what to look for. Their mother overturned large boulders, while the cubs copied her and picked up little stones. Shellfish and eels were on the menu, and the mother showed them how to catch crabs and not be nipped on the nose by their pincers. They had a lot to learn.

'The mother would sniff below a boulder,' Bertie noticed, 'and if she smelled a crab, she'd turn it over, and those boulders were really heavy. You and I would take two hands and a lot of effort to simply roll one over, but she just flicked them over with one paw. And, during the time we were there, we could see the cubs turning over increasingly larger stones.'

Above
The forested slopes of Vancouver Island and its small islets are parts of the vast ancient coastal, temperate rainforest of the Pacific Northwest.

Overleaf
The cubs are getting the hang of it! By watching their mother they have learned the skill of turning stones for what might be hiding underneath. Now all they need is practice.

The youngsters need plenty of protein for their growing bodies, and what takes less than an hour to find on the shore would take considerably longer to gather in the forest. Seafood is the single largest source of energy for these bears.

The cubs remain with their mother, learning the ways of island bears, until their second birthday; if they make it that far, that is. Male bears, which are twice the size of females, kill cubs, so black bear cubs – like many other young bears – must know how to climb and hide in trees. And, in these parts, there are also threats from other species of bears, and of course, people.

There are thought to be between 7,000 and 12,000 black bears on Vancouver Island and a few neighbouring islands, and, unlike many mainland bears, they are diurnal rather than nocturnal. This is probably because their cousin, the larger brown or grizzly bear, has been absent until recently, so they could be active during the day without fear of an attack. A few sub-adult grizzlies have island-hopped from the mainland and have taken up residence in the northern part of the island, but the black bears live mainly in the south so they've yet to be confronted by the more powerful grizzlies. Their main danger is human hunters. They account for about 700 bears per year, which, if the lower total population figure is correct, represents 10 per cent of the population. Biologists are concerned. They feel this take is too high for the long-term survival of the subspecies.

Below
At high tide, the water laps at the edge of the forest, and the bears have to wait for it to fall before they can go 'crabbing'.

Flower throat

After the snow melts, spring in the Rockies sees alpine meadows bursting into flower, and these blooms attract North America's smallest breeding bird and the world's smallest-bodied, long-distance migrant – the calliope hummingbird. It is a tiny thing, no heavier than a ping pong ball. Only the bee hummingbird, a non-breeding vagrant in the USA, is smaller.

The calliope hummingbird sits out the winter in the oak forests of southwest Mexico and neighbouring countries, and each spring it embarks on a journey of about 4,000 kilometres along the Pacific coast. It is one of the first migrants to head north, so it reaches the mountain meadows as soon as the first flowers appear. It then has to make best use of its time and find a mate quickly in order to complete its breeding cycle, for it has a relatively short window of opportunity at these high elevations and northern latitudes.

There are stacks of flowers, nectar and insects for fuel, and the male hummingbird has evolved an unusual way to attract females: he imitates a flower. At first he performs a dramatic U-shaped display, during which he flies up to about 30 metres, then dives down, producing a buzz from his tail feathers and giving a sharp call at the end of the dive. He then climbs again and repeats the display. If he attracts an audience, he hovers in front of the spellbound female with his magenta-coloured throat feathers fanning out just like the corolla of an alpine meadow flower. The receptive female responds by poking her bill at its centre, as if she was feeding… but surely she must find it odd that the 'flower' hovers about and even flies along with her!

Below and Opposite
Like most hummingbirds, the calliope hummingbird sips nectar from flowers and catches flying insects, but it also sips sap from holes created by sapsuckers.

Overleaf
The male hummingbird's throat adornment is a dead ringer for a mountain flower, and a great way to attract the attention of a female.

The pebble collector

In spring, many creatures put on an elaborate display to impress the opposite sex, and in many of North America's rivers there is a fish that does just that. It is the river chub, a type of minnow and one of the continent's most common freshwater fishes. When the time to breed approaches, the male of the species undergoes a distinct physical change. His head swells and white tubercles form between the eyes and the tip of the snout, and his body develops a pinkish-purple tinge. When the stream water temperature reaches 16–19°C, he seeks a suitable spot to make a nest, preferably where the current is slow and the water up to a metre deep. Using his mouth, he collects pebbles from the riverbed and builds a mound, not just any old mound, but one with a very definite structure.

First, he creates a shallow scrape in the streambed, and moves all the pebbles and stones to the edge of the hollow. Then, he starts to build. From as far away as 25 metres, he collects pebbles, some of them quite big, and erects a platform followed by a circular mound with a small trough in its upstream face. Director Sarah Whalley, floating face downward in the river (to the amusement of passers-by), observed this extraordinary feat of nest-building.

'Once a male got started there was no stopping him – one stone after another, and all the time watching out for females. There were times when he would put one stone on his mound and it would roll off, but he would immediately put it back in its rightful place. One day we watched a male build

Above
The male river chub is an avid collector of pebbles and stones. He collects so many stones that, during a typical lifespan of five years, he might shift several thousand individual stones in order to impress passing females.

a nest, then deconstruct it, before building it again about 30 centimetres to the left; they can be that particular.'

In all, the chub probably collects and carefully places more than 7,000 pebbles in a single mound. It can be up to a metre across, and this achieved by a fish that is just 14 centimetres long. The male then defends his nest against any intruding males. This is when the tubercles on his head come into play. When two fish slog it out, they attack with mouths open and bash the sides of their heads together. Generally, the interloper backs down and the resident male can look out for visiting females.

When a gravid female comes along and is impressed by his pebble nest, she enters the trough. He pushes her against the side and she deposits her eggs, which he fertilises with his milt. She does not drop all her eggs in one nest, but distributes them amongst several other nests in the stream. She then swims off and leaves him to fan the eggs to ensure they are well oxygenated and safe from nest predators.

The male river chub is often not alone. Nest associates join him; usually other species of even smaller fish that take advantage of his building skills and deposit their own eggs and milt around the mound. Some, such as the blackside dace, are so dependent on his nest mound and the presence of the fish itself that they cannot reproduce unless stimulated by his milt. This relatively small fish has a big impact on the life of its river.

Above
The male river chub's building site can be quite extensive, with conical piles of coarse gravel and pebbles more than a metre across. The largest pebbles might weigh more than 200 grams and are placed on the outside, but most are about 100 grams. Many are from rock strata not common in the neighbourhood, so the river will have washed them down.

Living lights

Look up at the night sky in the wilderness and, with so little light pollution, the heavens glow brightly with the stars of the Milky Way. Look down at the ground in Mississippi and the stars appear to flash in synchrony... except they're not stars: they're fireflies, also known as lightning bugs, and their breeding time is two to three weeks in early summer. This is when males give up flashing at random and join together to flash in synchrony. The male Mississippi fireflies from the genus *Photinus* flash quick single bursts every two seconds, the duration of the flash and the gap between flashes being different for the three species of synchronous fireflies that live in North America.

They begin their light show just after sunset and flash away for about an hour. Some have another bout just before dawn. Males do not join in for the entire time. An individual might flash with the others, then stop for a while, before joining in again. The female firefly watches from a distance and then flashes a coded reply. She is looking for the best flashers, and sets up a 'photic-dialogue' to determine which one she will take as a partner. The males then crowd around her, each trying to mate with her, but she does not accept the first male to reach her. She is more discerning, showing avoidance behaviour to ensure she mates with the best of the bunch.

Despite their common name, fireflies are not flies but small beetles with light organs on the underside of their abdomen. Most adults do not feed and are short-lived, but the females of larger *Photuris* genus are aggressive predators, which are not only out to trap a meal, but also to sequester defence chemicals from other species of fireflies. The smaller *Photinus* fireflies, for example, manufacture lucibufagins, which protect them from ants, jumping spiders and even birds, and the large female *Photuris* firefly gains immunity by feeding on them, especially the males. She catches them in mid air, a behaviour known as 'hawking', but she might also use subterfuge

The female *Photuris* is known popularly as the 'femme fatale' firefly. She mimics the smaller female *Photinus*'s 'come hither' flashes in response to the signals from male *Photinus* fireflies, so instead of meeting the firefly of their dreams, the males' courtship turns into a nightmare: they're grabbed and eaten. And these femme fatales have an even craftier trick up their sleeves to get fresh meat.

The large female carefully positions herself behind a spider's web. The males come flying in and blunder into the sticky strands. They're trapped. The spider wraps them up in silk, to be liquidised and supped up later, but the male fireflies continue to pulse, even inside their cocoon, which serves to attract more into the web. The *Photuris* femme fatale is big enough to fight off the spider. She cuts the food packages out of the web, and consumes them herself. She is not only a ferocious predator, but also a brilliant flash mimic and a crafty kleptoparasite.

Opposite
A firefly blunders into a spider's web, and while the spider is wrapping it up in silk, the beetle continues to glow, attracting even more fireflies into the web.

Overleaf
Like the tiny incumbents of a fairy glen, synchronous fireflies display in an extraordinary symphony of light.

Towns under siege

On the eastern side of the Rockies, to the west of the Mississippi River, is an extensive area of prairie, steppe and grassland known as the Great Plains. One of its most famous inhabitants is the prairie dog, not a canid but a rodent. It digs extensive tunnel systems, where it can escape from predators and the vagaries of the weather.

Prairie dogs tend to live at altitudes between 600 and 3,000 metres above sea level where the summer air temperature can reach 38°C; in winter it can plummet to minus 38°C. Add to that the dangers from hailstorms, blizzards, floods, tornadoes and wildfires, and the burrow becomes a vital refuge. It can have up to six entrances. Some are simple holes in the ground, while a pile of soil surrounds others that can be used as lookout stations. They are also part of the air-conditioning system, for they help circulate the air in the prairie dog home. Fresh air enters the flat tunnel entrances, and leaves via the chimney-like ones.

Inside the tunnels are entire families. The family is the basic prairie dog unit, though generally a number of families live close together to form a 'town', sometimes covering several hectares. Each tunnel system is divided into chambers – shallow chambers for sleeping and deeper chambers used as nurseries. They are so desirable that other creatures try to muscle in, and occasionally the town is under siege from a surprising quarter – burrowing owls. It seems these diminutive owls do not do a lot of 'burrowing'; instead, they dive-bomb the prairie dog families in order to evict them.

Larger predators are a more serious problem, although the prairie dogs have an effective warning system. The mound beside a tunnel entrance affords a good view of the surrounding grassland, and they have good long-distance dichromatic colour vision. This is a form of colour blindness in which the animals can recognise two of the three primary colours, so that green and yellow things, such as grass, cannot be told apart, but anything red or blue is seen as different from green and yellow.

When a predator appears, the sentinel gives a warning cry. It is thought that they have specific warning calls for different predators, with an indication of what the animal should do. A single call, for example, indicates how quickly a predator is approaching and whether it is a perceived threat, so the animals know where to look for the danger and what to do to escape it, or whether to simply ignore it.

If a hawk is spotted, all the prairie dogs in its flight path dive into their burrows, while those nearby simply watch what the hawk is doing. Coyotes

Below
When an alert is sounded, all the citizens of the prairie town adopt curiously human postures to check out the danger.

Opposite
Prairie dogs stand a lot, even when having lunch. It is an effective way of spotting danger approaching and communicating with the rest of the gang.

Left
When danger is spotted, prairie dogs engage in a series of contagious moves, which see them all throwing back their heads and shouting loudly.

Below
The American badger is one of the most dangerous natural predators of prairie dogs, but fortunately burrowing owls have a problem with them too, and they don't hold back in coming forward.

Overleaf
Monument Valley is the quintessential image of the American West, due in part to the feature films of director John Ford.

and domestic dogs prompt the entire family to emerge from its burrow. From here, the prairie dogs watch every move the predators make. Humans elicit the opposite reaction: the prairie dogs all dive for cover. The fact that they can 'talk' about specific dangers and particular colours indicates prairie dogs have one of the most sophisticated animal communication systems that has been understood so far.

While hawks and coyotes are relatively easy to avoid by escaping into the tunnels, there is one creature that is feared above all others – the American badger. In spring and summer, when the grass has grown long, the badger can sneak up on a family undetected. Most vulnerable are young prairie dogs in the nursery, for the badger can enter the burrow, dig if it has to, and take what it wants. In this situation, burrowing owls can turn out to be good neighbours after all: they mob badgers and drive them away.

Meep, meep – it's the real roadrunner

The deserts of Southwest USA are home to an unusual and rather engaging bird – the aptly named roadrunner. It is a member of the cuckoo family that tends to run through Arizona's arid regions rather than taking to its skies. The roadrunner is a predator, but instead of flying and dropping onto victims, like most predatory birds, its long legs enable it to run them down, whizzing along at speeds up to 20 mph. Strong feet make short work of a lizard or even a scorpion, and its movements are so fast that it is one of the few animals that can kill and eat rattlesnakes. A pair of birds might work together, one distracting the snake, while the other pins down its head and delivers the coup de grâce with its outsized bill.

It is well adapted to its arid surroundings. It doesn't need to drink, for it gets all the moisture it needs from its prey. Excess salt is secreted from glands close to the eyes, an anatomical feature more usually associated with seabirds, but it shows just how well adapted this bird is to desert life.

For the filmmaker, finding one can sometimes be a problem, for the bird's X-shaped footmarks can be confusing. It looks as if the bird is running in opposite directions. However, often as not, the birds will find you, as field director Jo Avery discovered.

'These zany birds really live up to their Loony Tunes namesake. As the film crew stepped out of the camera car to film the birds, the birds stepped into the driving seat and sat watching the crew. They were real characters!'

Below
Roadrunners hunt on the ground. They are so attuned to their environment and so swift that they can catch a fast-running lizard with ease.

Opposite
The roadrunner or chaparral bird is about 24 centimetres long. It has a head crest, and an oversized black bill. After cold desert nights, when it lowers its body temperature to conserve energy, it exposes dark patches of skin to the sun in the morning in order to warm up to an operating temperature.

Overleaf
The deserts of Southwest USA are prone to brief but exceptional electrical storms accompanied by torrential rain.

Mullet run

With the onset of autumn, many animals are on the move, migrating to places where they optimise their chances of feeding, procreating or simply staying alive. All across North America, millions of bird migrants join traditional flyways, all flying south to avoid the cold weather and poor pickings. Some travel down the East Coast of North America. It's a familiar sight, but less well known is that in the water below them are fish doing exactly the same thing.

As the temperature drops, striped grey mullet leave their summer haunts off the Carolinas and Georgia and head south. Most are just offshore, within 100 metres of the shoreline, while others enter inlets and follow the intra-coastal waterway. By late August, the first shoals will have reached the central East Coast beaches of Florida, with a peak in numbers between mid-September and the end of October. The shoals are so enormous, that they look like the shadows that dark clouds make on the water, and they might darken coastal waters for as far as the eye can see. On a calm day, almost imperceptible ripples on the surface, known as 'nervous water', give them away, but a sure-fire sign that the mullet have arrived is the presence of gangs of excited seabirds – pelicans, cormorants and gulls – over the ocean.

Above left
An enormous tarpon leaps from the water in pursuit of mullet.

Overleaf
Like some gigantic oil slick, the mullet crowd into Florida's inshore waters, the drama below revealed by the occasional splashes of white water above.

The birds take their cue from the big predators. Large metre-long tarpon and snook, along with blacktip and spinner sharks, move inshore for the feast. The shiny silver flanks of the tarpon are a giveaway, and the mullet can take evasive action, an entire shoal moving as one; but, if the tarpon turns on its side, the smaller fish cannot see the predator approaching. Where it attacks, the water boils, as mullet explode en masse from the surface, leaping up to a metre out of the water to escape its jaws. The tarpon and sharks follow, punching up through the surface and twisting in the air.

When the mullet reach the southern tip of Florida, sexually mature fish that have escaped the predators' jaws swim out to the warm waters of the Gulf Stream, where they spawn. The fertilised eggs catch a ride in the northward-flowing current, and, some time later, the tiny juvenile fish head back inshore to shallow coastal waters and estuaries, and the 'mullet run' cycle starts all over again.

Above
Mullet explode from the sea's surface in a desperate attempt to avoid the predators below

Mermaid springs

In autumn, when the trees are turning, all thoughts are of New England and the spectacular colours of the leaves at this time of year, but surprisingly this colour change can be seen as far south as the southern swamps. At Caddo Lake, on the border between Texas and Louisiana, the flooded bald cypress forest is a magical place with trees draped in Spanish moss, and they are very old. One tree is estimated to be at least 2,624 years old, making it the fifth-oldest-known tree on Earth. In the autumn the leaves all change colour, turning shades of red, brown and yellow. It is a direct result of the continent's geography.

With no east–west mountains to stop it, the cold Arctic air pushes down to the Gulf coast. During particularly vicious cold snaps, alligators are sometimes seen with their eyes and nostrils protruding from any ice that has formed on the surface of the water. It is a climate regime unique to North America, and one creature has had to take measures to avoid the chill or die.

The Florida manatee is a subspecies of the West Indian manatee, a fully aquatic mammal that is at home in the tropics and subtropics. If the water temperature should drop below 20°C, it could mean the deaths of both adults and babies because they have little insulating subcutaneous fat to keep them warm and a very low rate of metabolism. Their answer is to seek out areas fed by warm springs, where over a hundred manatees can be seen huddled together, their bodies being cleared of algae by local fish. Like some slow-motion barn dance, all the manatees float as one to the surface, take a breath, and then sink back down again. Sometimes, they even link flippers. One or two savvy alligators might join them, and there is a temporary cessation of hostilities. The two species, which make for unusual bedfellows, lie alongside each other in the warm water.

The sanctuary, however, can be short-lived. With so many animals together, their food supply – mainly aquatic plants – is quickly exhausted. The manatees must then brave the cold waters to find fresh supplies. The problem is that below 20°C their gut shuts down. The result is that many die from thermal shock, the cause of a substantial number of manatee deaths in some years. A few seek out the warm water outlets from power stations, but, with plants shutting down, that source of warmth is less available. The US Fish and Wildlife Service is now trying to work out how to provide more warm water for more manatees.

Above and Opposite
Florida manatees crowd around natural hot springs during winter. It is the only way they escape water temperatures that might kill them.

Previous page
Caddo Lake is a surrealist's dream: a flooded cypress forest on the border of Louisiana and Texas.

Behind the Scenes

It's no picnic at Hanging Rock

Hudson Bay is a treacherous place to be at the best of times. For much of the year, it is covered by ice, but during the few ice-free months of summer, the huge tidal range sees powerful currents that, faced with fierce oncoming winds, drive up steep and violent standing waves and generate swirling whirlpools that can sink a small boat in seconds.

The Bay itself is enormous, about three times the size of the British Isles, and on its western shore, about 45 kilometres north of Churchill – the self-proclaimed polar bear capital of the world – is the mouth of the Seal River. It is a magnet for belugas or white whales, over 3,000 of them. They come here to slough off their old skin by rubbing against the shingle on the riverbed. It is also the place where mothers bring their calves and even give birth, for the river water is several degrees warmer than in the rest of the Bay, the perfect beluga nursery. To film them underwater, though, producer Chadden Hunter's team needed a bit of ingenuity, as cameraman Bertie Gregory found out.

Above
A group of inquisitive belugas form a nervous but orderly line when checking out the camera boat.

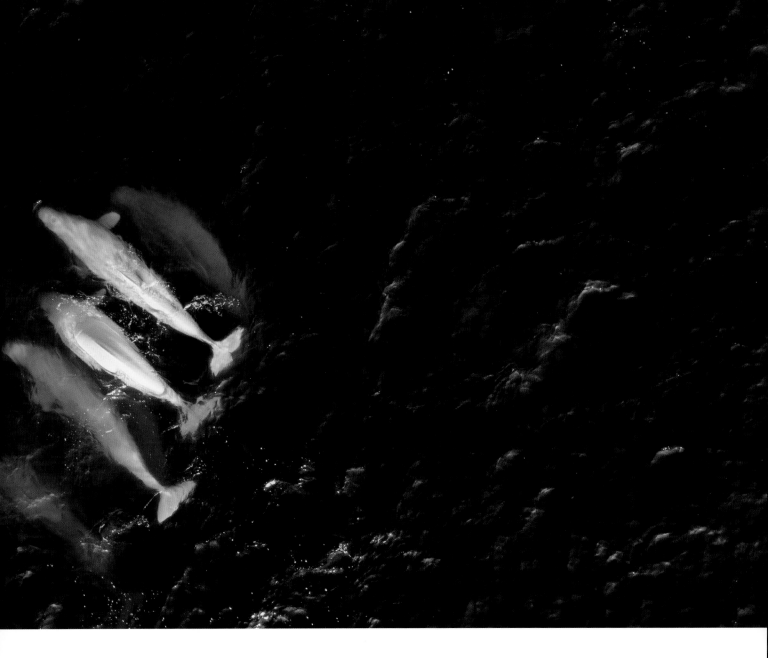

'The belugas were wary of a stationary boat and they kept their distance, but they seemed to like the bubbles coming from the propeller, so I was pulled on a rope behind the boat. The belugas loved it. They would swim alongside me, and even bumped the camera with their melons. They also became more curious if you sang, so there I was being dragged around Hudson Bay in icy Arctic water while singing the theme tune to *Jurassic Park*!'

The beluga nursery, however, is not entirely perfect. Those famous polar bears are here too, and they seem smarter than the average bear.

The Bay is scoured by one of the largest tidal regimes in the world. When the tide is out, it really is out. An enormous strip of mud, sandbanks and shallow waters extends 10 kilometres from the shore, so on its return the incoming tide is moving at the speed of horses. The whales ride in with the tide, but the polar bears know this too. They have a number of rocks at key vantage points on which they simply sit and wait, sometimes for hours on end. Showing

extraordinary patience, a bear will wait for the moment a whale swims just close enough, and then, with all its strength, leaps right on top of the beluga, an incredible sight! It was a scenario that became immediately clear to wildlife cameraman John Shier, the moment the team overflew the location.

'Hudson Bay is relatively shallow and its bottom gradually descends. This means that immense boulders can be over a mile from shore at high tide and still have their tops above the water. Rarely do you quickly figure out the mechanics of what exactly is going on when you're trying to film animals hunting, but on this shoot all it took was flying over the shore for a few minutes to piece it together. It was immediately clear that at high tide the tops of those giant boulders, standing in 2–3 metres of water, provided the polar bears with the perfect platform for ambushing their prey.'

Having caught a whale, the bears will bite and wrestle it into shallow water, no mean feat considering belugas are twice the weight of the bears. Here, the whale's flesh is pulled away in neat forensic strips, a meal that is often shared with several less successful hunters. John thought it was unusual behaviour for the normally solitary polar bear.

'I was really amazed to see how many bears showed up every time a carcass was brought ashore. At first, I thought it was a few large, dominant male bears. This was an easy hunting opportunity for them, but younger bears and mother bears with small cubs would come running out of the bushes and down the beach every time a successful hunt occurred. I realised that while only a few bears may be doing the hunting, a whole community of bears was spending their summer on this stretch of shoreline, drawn here by the reliable source of calories.'

Above left
On his chosen rock, a polar bear prepares to jump onto the back of a beluga.

Above
A successful bear turns the sea red as he butchers his catch.

It was all of this extraordinary behaviour that the production team was here to film, but it wasn't easy sailing. They were filming with drones and an extremely large camera rig from a small boat, so they were open to the vagaries of the weather and the dangers of working with North America's largest terrestrial predator.

Chadden describes it as an energising place to work.

'Normally, you try to be invisible, just observing your subject, but these bears just sought you out, especially the young males, and they come right up to you; but they are nothing compared to the weather and sea conditions.

'There is zero backup. No lifeboats, no rescue craft. The tides are fast and violent, so it's difficult to tell where you are, and, to top it all, thunderstorms sweep in, the lightning striking the water all around you. You feel extremely vulnerable in an aluminium boat with a huge metal camera boom sticking up, the tallest thing for miles around. Fortunately, our captain was very cautious, but even he wasn't prepared for what happened.

'A young male bear swam over to us and checked us out. "Are they edible?" he seemed to be thinking; but this one did the strangest thing. He grabbed the anchor line, lifted the anchor, and was pulling the boat like a fisherman reeling in his catch. Unfortunately, he caused the boat to drift onto a rock – just the sort of place that polar bears like to ambush belugas. We were really in trouble. It was serious enough being stranded, but as the tide went out we were in danger of the entire boat tipping over.

'There was nothing else for it. The bear had had his fun and had moved away, although it was still nearby, so John stripped off and got into the water,

Above
Normally solitary, the hostility between bears breaks down when there is a surfeit of food to be had.

where he was able to gently lever the boat free from its rocky pinnacle before the tide dropped completely. All the while we watched his back for hungry bears.'

'With me under the bow pushing and with our skipper pulling on the stern,' recalls John, 'we finally slid off the rock and were free again, and just in time as we barely had enough water left to weave our way through the rapidly emerging boulders to deeper water. The funny part is the polar bear that had been playing with our anchor was still around the whole time. It was only after I climbed back aboard that it dawned on me: I had gone swimming with a bear!'

Although the bears had taken an immediate interest in humans, they were a little slower at catching belugas. From the drone, Bertie could see that young bears would jump too soon or too late, and miss the catch, while the more mature individuals would wait for hours before making a move.

'We spent many days just watching bears perched on rocks, with belugas swimming agonisingly close to them. I remember one time looking down the viewfinder and seeing whales surfacing to breathe right in front of the bear, and I was shouting, "Jump!" Eventually he did, but it was very sad to see

Above
Polar bears and camera crew alike wait patiently for the belugas to return on the incoming tide, although a singing, swimming cameraman (3rd top right) had them foxed!

such a magnificent animal being killed. To see the planning, patience and intelligence of the polar bear, followed by the power needed to kill and drag an animal far heavier than itself, was quite something.'

It was also intriguing behaviour, and John was convinced that many of the bears are getting more calories in summer than they are at any other time of the year.

'Some of the bigger bear's bellies nearly drag on the ground and their whole body jiggles with fat, so these bears may actually be at their fattest in August, something unheard of for polar bears. The meat and blubber the bears are getting from the whales is helping these bears keep weight on throughout the summer. As our warming climate makes the Arctic warmer and more ice-free, polar bears are facing a future of increasingly shorter seal-hunting seasons, which for them, isn't much of a future at all; but at Seal River, and at other places like it, we see, perhaps, a bit of hope for some small populations of bears that are intelligent enough to have figured out how to use the landscape to access the food they need to survive.'

Below
The belugas are in their element in the sea, and even a good swimmer like a polar bear, considered by biologists to be a competent marine mammal, fails to scare them.

Antarctica

Producer: Fredi Devas

Introduction

With 24 hours of darkness in the depths of winter, Antarctica is the coldest, windiest, and, surprisingly with all that ice, the driest continent on Earth. With an annual precipitation of less than 100 millimetres at the South Pole, most of Antarctica is technically a desert. It is also unbelievably cold. The air temperature can plummet to minus 89.4°C, as it did in July 1983 at Russia's Vostok Station. It was the coldest natural air temperature ever recorded in a conventional way. However, taking data from various sources, including readings from satellites, scientists believe the coldest parts of Antarctica are on the Eastern Antarctic Plateau, where there are no weather stations and the snow surface temperature regularly reaches minus 98°C during the polar winter night, making it the coldest place on Earth.

Ice defines the white continent and the ocean that surrounds it, so it is a challenging place for life to survive in. Living organisms have two choices: they either try to avoid it by moving north, and find themselves on subantarctic islands where they compete with many millions of others for food and space, or they attempt to embrace it by remaining south, where they find solitude but must endure one of the most hostile places on the planet.

There are few native terrestrial species on the mainland, most no bigger than midges, springtails and mites. On the coast and on subantarctic islands – forming the wider Antarctic region, however, the picture is very different. The Southern Ocean and the southern extremities of the Atlantic, Pacific and Indian oceans are packed with marine life, enabling seals and penguins to live here all year round. Great whales – including blue, fin, right, humpback, minke and sperm – and countless seabirds, including several species of albatross, arrive here for the summer, when most animal communities depend on a shrimp-like crustacean that is probably the world's most numerous animal – krill.

Right
Wiencke Island, the southernmost major island in the Palmer Archipelago, is reflected in the calm waters of the Gerlache Strait, close by the Antarctic Peninsula.

World's southernmost mammal

Weddell seals embrace ice, rather than avoid it. They especially like immobile sea ice. For them, it is a relatively safe place to be. In spring and summer, they haul out onto the surface of the ice to rest, moult and give birth, and, with so much of it around them, they are out of range of Antarctica's other large predators – killer whales and leopard seals. That safety comes at a price.

The seal is a large animal, up to three metres long, with a surprisingly small head, but its lithe body enables it to dive down to 720 metres for 45 minutes while chasing and catching large benthic (bottom-dwelling) or midwater fish,

Below
A young Weddell seal, like its parent, has a thick layer of blubber below the skin and a thick fur coat. The large eyes are an adaptation for hunting underwater at low light levels.

such as Antarctic cod. It has even been seen to blow bubbles into cracks in the ice to flush out fish hiding there. It also takes squid and octopus, and the occasional penguin.

As an air-breather, however, it must return to the surface to take a gasp of air. In the sea ice, this means finding relatively small breathing holes that the seal itself has excavated. It creates several, which are closely guarded, and it rarely travels far from them. It will venture up to 20 kilometres from its territory to feed, but it always returns to the same breathing holes, no mean feat in the darkness of midwinter.

Mature male seals tend to set up territories around the breathing holes of several females, and they defend the area with a barrage of very loud, alien-sounding calls and occasional fights. After the previous year's pups have been weaned, usually about mid-December, mating takes place in the water, although this has only been observed once. Embryo implantation is delayed, so a single pup is born on the ice in spring; all well and good, but this is the Antarctic. It can be a bit of a shock sliding from its mother's body temperature of plus 37°C to an air temperature of minus 25°C in just a few seconds.

Mother provides it with fat-rich milk. It has up to 60 per cent fat, 10 per cent protein and little sugar, so the pup gains almost two kilograms in weight

Below
In winter, the seal must keep breathing holes open using its teeth to rasp away the ice.

each day. It means mother will have lost a lot of weight and needs to feed, so she will leave her offspring. While lying on top of the ice, the pup is very exposed, although its mother usually returns before the weather turns bad. She bends her body around her pup, and protects it as best as she can. Even so, dead pups can be found on the ice because mother didn't get back in time.

When learning to swim, it must haul itself in and out of those breathing holes; and it must learn fast for it is weaned and must fend for itself within six weeks. Newly weaned youngsters and sub-adults tend to move to the fragmented pack ice, in order to stay clear of the adult territories on the fast ice. Here, they are vulnerable to attacks by predators, especially killer whales.

The orcas work together, sometimes seven abreast, to create a wave that tips them off ice floes and into the water, and they have a distinct preference

Opposite top
A young seal learns to swim and to haul itself in and out of the breathing hole from about one week old.

Opposite bottom
Orcas or killer whales target young Weddell seals, which they tip off ice floes. Adult seals have their territories amongst thicker, less broken ice, away from these apex predators.

for Weddell seals, even though they make up only 15 per cent of the Antarctic seal population. They ignore the more abundant crabeater seals; perhaps the crabeater's bigger teeth and more aggressive attitude scare off the predators, but nobody really knows why. Any Weddell seals that survive the onslaught will not return to the fast ice and the area where they were born until they themselves are ready to mate.

In summer, breathing holes generally remain open or the fast ice breaks up with large gaps between the ice floes, but in winter, the holes constantly freeze. The seal keeps them open using its canine and incisor teeth. It rasps away at the ice, but this wears them down. Eventually they are worn so badly that the seal is unable to catch and hold prey or it can no longer keep the breathing hole open. It either starves to death or drowns.

By working with the ice, rather than avoiding it, the seal is able to live in Antarctica all year round, but it will die when about 20 years old, half the age reached by other Antarctic seals. It's a high price to pay for solitude.

Below
Each hair in a young seal's fur has an outer surface composed of overlapping non-living cells. Hairs with large, smooth-edged cells feel silky.

Overleaf
Under the ice is a strange labyrinthine world linked to the outside by several vital breathing holes. The seal can find its way about even in the darkness of winter.

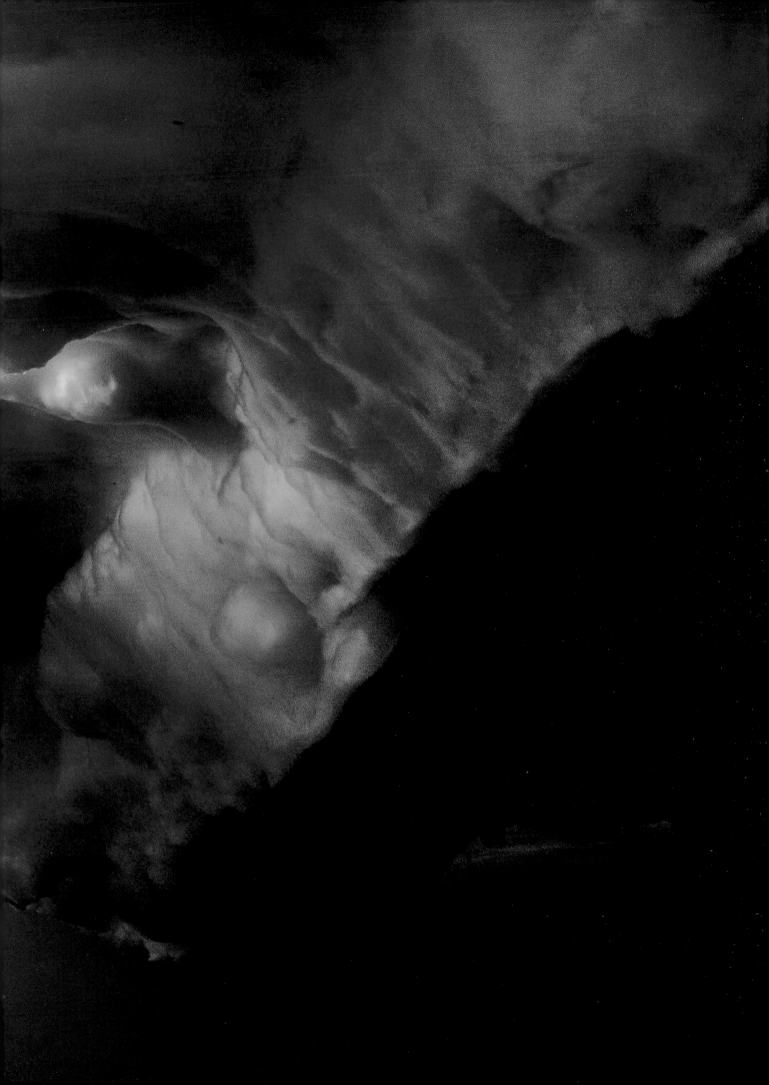

Close to the Pole

Spring in the Antarctic must be a welcome relief from the long, dark winter. While much of its wildlife will have relocated to slightly warmer locations with ice-free seas during the worst of the weather, the shortening nights and slight rise in temperature trigger the big return, and living and breeding closest to the South Pole is a relatively small and seemingly delicate bird – the snow petrel. It is about the size of a dove, with long wings. Its outer feathers are pure white, its eyes dark brown, its bill blue-black, and its body is kept warm by sub-dermal fat and a thick insulating layer of underdown. Some observers have described it as the most beautiful bird in the Antarctic. It is also the most abundant bird on the mainland.

The snow petrel breeds the furthest south of any bird, aside from its arch-enemy, the south polar skua. Many nests are inland on isolated rocky peaks and ridges, known as nunataks or glacial islands, which protrude from Antarctica's ice cap. Here, surrounded by ice, the birds can be up to 440 kilometres from the sea, like those that nest on the slopes of the Theron Mountains on the

Below
Snow petrels get together at breeding time during the southern spring, when the male has to convince the female that he is fit enough to be her partner.

eastern side of the Filchner Ice Shelf, the most southerly known breeding site
of petrels and skuas. Their regular commute to and from the nest site might
seem rather excessive just to get provisions, but they nest so far from the sea
because bare rock is at a premium on mainland Antarctica.

The nest itself is a modest scrape on the ground lined with pebbles, the
best sites located where rocky overhangs or rock crevices protect the birds
from the worst of the weather, but these are highly sought after and often
fought over. The birds screech loudly, clash bills, grapple with their feet, flap
their wings violently, and, if all that fails to see off an intruder, they regurgitate
foul-smelling, orange-coloured, waxy stomach oil, and *that* is sure to work.
Any birds with their white feathers splashed orange must have been at the
receiving end of another's wrath. They can try snow bathing to remove it but,
if they are splashed on the wings and cannot remove the gunk, they cannot fly
so they inevitably die. Likewise, if a bird vomits too often, the loss of food, and
therefore that source of energy, means that later in the annual cycle it would
be forced to abandon its eggs in order to go to sea to feed.

Snow petrels mate for life, so when the breeding season of the Southern
Hemisphere begins the birds re-establish their pair bonds. They perform a
courtship manoeuvre that sees the female flying headlong towards the cliff, only
to turn at the last moment. The male must follow close behind, not knowing

which way she will turn, and, if he succeeds in matching her every twist and turn and does not plough into the rocks, he has passed the test. The birds then drop down onto the nest site, where they raise their heads, fence, and mutually preen. Both parents take turns to incubate their large single egg and care for the chick, but the youngsters are highly vulnerable to predation by skuas, another reason that covered nest sites are at a premium.

The adult petrels themselves look very delicate in their snowy-white plumage, but underneath that gentle facade they must be tough to survive in such a demanding environment – perhaps the toughest birds on the planet. In flight, they glide less and flap their wings more than other petrels, flying low over water, but high over land to avoid skuas. While away from the nest a parent might travel hundreds of kilometres to fetch food for their chick, which can be krill, fish or squid, and they are not averse to tucking into some carrion from a dead seal or penguin.

There are thought to be more than four million snow petrels, but for how long is anyone's guess. Climate change could be threatening their very existence. A reduction in sea ice might mean a shorter trip to the sea but, when they get there, they could be confronted by a food shortage because krill depend on sea ice, and there is less of it than there used to be in areas where most krill are found.

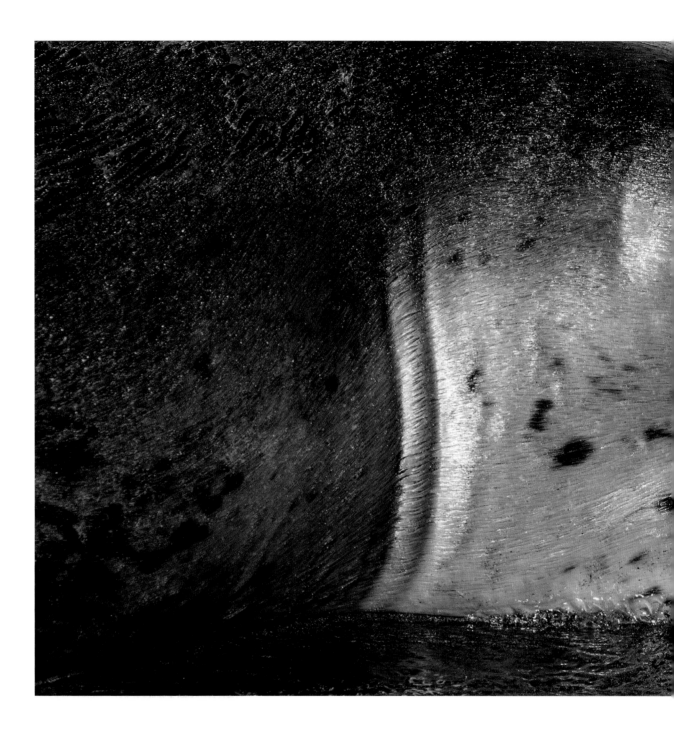

The assassin's smile

The leopard seal often hides amongst ice floes, where it can be the bane of penguins. It is a wily predator – the 'fox' of Antarctica – ostensibly happy-go-lucky and often clearly visible just offshore, but quite capable of sneaking below and hijacking any penguin either entering the water or trying to leave it. It is commonest in the western Antarctic, where leopard seals patrol the pack ice.

Sharing their living space here are gentoo penguins, the third largest, after emperors and kings. Fledglings are especially vulnerable. When they take their first swim, they are naive and slow, and they are obstructed by the brash sea ice that forms in autumn, an obstacle between them and the safety

of the open sea. It means youngsters fail to swim properly and they cannot see what is lurking below. Leopard seals seize them, smash them against the sea's surface until dead, and then flail them from side to side in order to rip them into manageable pieces. Watching from a safe distance was executive producer Jonny Keeling.

'I think the most incredible part of this story is that the young penguins' first ever swim and their first encounter with the ocean involves navigating crushing ice, big waves and trying to avoid a vicious predator, and all without any assistance from their parents!'

Above
The leopard seal has large canines for grasping, but also sharp and complex molars used for straining krill from the water.

The predator itself can be up to 3.5 metres long and has a massive reptile-like head, with a sinister assassin's smile. The large mouth has sharp 2.5-centimetre-long canines, and its back teeth interlock. They can be used to sift krill from the water. Young seals, more than adults, tend to eat krill, along with fish and squid. When the animals mature they switch to catching penguins and other seals.

They patrol the shore and the edge of the ice, and, as winter arrives, there is a big change in the extent of that sea ice. The ice grows outwards at a rate of about three kilometres a day, which means by mid-winter Antarctica will have doubled in size. Underneath, it is another world entirely.

Above
A large group of Adele penguins looks nervously out to sea, searching for any telltale signs of leopard seals or orcas.

Opposite
Leopard seals are less agile on land so, if the penguin can make it to the rocks, it has a chance to escape. If not, it is caught and flayed alive.

Icy underwater Narnia

The southernmost navigable body of water on the planet is Antarctica's McMurdo Sound, except that in winter it freezes over with sea ice three metres thick. At the surface, it is very much a black-and-white world, but venture below and a magical underwater Narnia spreads out before you.

Close to the shore, the sea is rarely free of ice and, while the air temperature fluctuates greatly between summer and winter, Antarctic bottom water remains, on average, a steady minus 0.8°C to minus 2°C all year round, depending on the location, because saltwater freezes at a lower temperature than fresh. It is one of the most stable environments on the continent, and it is chock-full of life.

The sea floor is an extraordinary riot of colour – carpets of sea anemones with wispy tentacles, gaudy pink, purple and yellow starfish, bright yellow cacti and green globe sponges, red sea urchins, creamy sea cucumbers, pink and flimsy featherstars and spiky silver-coloured brittlestars. There are translucent tunicates, jewel-like cup corals, countless species of marine worms and cold-adapted octopuses, along with jellyfish that carry amphipod passengers. It is a secret kingdom dominated by the marine invertebrates, and some of them are giants.

Above
Invertebrates are the dominant life forms below the ice. Algae grow on the underside of the ice, food for krill.

Giant Antarctic sun stars, nicknamed 'death stars', have about 50 arms and are 60 centimetres across. They catch krill with tiny tooth-lined claws or pedicellariae on their curled-up arms. Giant sea spiders can have a leg span of 40 centimetres; giant isopods, marine relatives of woodlice, are 10 centimetres long; and ribbon worms are more than two metres long. Sponges are two or three metres tall and are thought to be at least 100 years old. One species – the giant volcanic sponge – is thought to have reached the grand old age of 15,000 years, making it one of the oldest living animals on the planet. All these creatures grow exceptionally large and have extremely long life spans, because metabolic rates are low and oxygen levels in cold water are high.

Down here, the invertebrates dominate, and fight amongst themselves. Sea anemones catch jellyfish, grabbing them by their tentacles and hauling them in. Soft corals here break all the rules and can bend over to gather food close to the seabed; and they all, to some extent, rely on the ubiquitous swarms of Antarctic krill, both living and dead. In summer, their moulted skins and dead bodies, along with debris from the phytoplankton and faeces from whales form dense clouds of 'marine snow' that trickle down from above to feed much of the Antarctic's benthic life.

The krill crisis

Antarctic krill are shrimp-like crustaceans about six centimetres long, but what they lack in size, they more than make up for in numbers. It is estimated that their combined weight, at any one time, is more than the total weight of the world's human population. They are probably the most numerous animal species on the planet.

Antarctic krill are also a keystone species. Almost every animal living in the Antarctic depends on them. Even if it does not feed directly on krill, it will probably catch another that has, but our warming world could have an unforeseen impact on their populations.

During winter, krill migrate into the deep, and the sea ice traps substantial quantities of algae. When it melts, the algae bloom but, because the ice is melting earlier than usual, the krill are still deep down. If they do rise to the surface, they are in a slow metabolic state, and so they feed inefficiently. It means the algae die and sink to the seabed before the krill have had a chance to feed properly – a trophic mismatch.

Scientists have also noticed that most of the krill they see are generally longer and older adults. There are fewer juveniles, so recruitment is substantially less than it once was. Add this factor to the early ice melt and we have a potential serious population slump for every species relying on krill.

A third problem is that people are catching krill in huge quantities. Nowadays, fishing boats equipped with suction devises harvest more krill than they did previously, much of it as food for livestock or satisfying a demand for krill-based health products containing omega-3 fatty acids. For the moment, though, krill still gather together in astronomical numbers. A single swarm can have up to 30,000 animals per cubic metre, which makes them a primary target for baleen whales.

Above
Hundreds of fin whales, probably the largest aggregation seen in recent times, are feeding on swarms of krill close to Elephant Island in the Southern Ocean.

Opposite
Krill is a shrimp-like crustacean and a keystone species in the Antarctic. It feeds almost everything – whales, seals, penguins and other seabirds, fish and many types of invertebrates.

The whales are coming!

The western Antarctic sector of the Southern Ocean is a magnet for whales in summer. Over 3,000 humpback whales feed in the coastal waters of Bransfield Strait, between the South Shetland Islands and the Antarctic Peninsula, and almost 5,000 fin whales feed along the edge of the continental shelf in the Drake Passage. While both feed on Antarctic krill, the fin whales also catch another species of krill – *Thysanoessa macrura*, a close relative of the Arctic krill. The production team witnessed hundreds of fin whales feeding on a krill swarm, probably the largest aggregation of great whales that has ever been filmed.

Humpbacks concentrate the krill even more by bubblenet feeding, while blue whales must have the biggest maw in the animal kingdom, but they only open their mouths when the concentration of krill makes it worth their while. All these whales are filter feeders. They engulf a huge volume of water containing krill, and then, using their huge tongues, squeeze out the water through their brush-like baleen plates, trapping a great ball of krill that the whale swallows.

Populations of most of these whales are bouncing back after the slaughter during the late nineteenth and early twentieth centuries. One of the first to be hit badly was the southern right whale. It was inquisitive, approaching closely to ships, so it was easy to harpoon, and when it died it floated. It was the 'right whale to catch', hence its name, and now it is back.

Above
Southern right whales are easily distinguished from other baleen whales by the white callosities on and around the head. These are calcified skin patches infested with whale lice, barnacles and parasitic worms. Their purpose is unknown, although whales have been seen to scratch each other using them, and it's thought they could be used against predators, such as orcas.

Some southern right whale populations have doubled during the past decade, and each year they migrate between feeding and breeding grounds. They feast on krill and copepods down south in summer and, while many head off to South America and South Australia to mate and give birth during the winter months, a few remain in the Antarctic region. They seek the shelter of subantarctic islands, such as the Auckland Islands.

The population here is making a remarkable recovery. Not that long ago as few as 35 breeding females survived, but now the population is thought to exceed 1,000 and, despite the horrors of the recent past, they are still amazingly inquisitive, following divers around like 14-metre-long puppies. All age groups are present. Cow and calf pairs tend to remain in shallow waters, bulls further offshore. Mating adults and calving females appear in the natural harbour of Port Ross at the north end of the island. While they interbreed with other New Zealand right whales, they appear to be isolated from most other right whale populations. Genetic diversity is low, showing just how perilously close they came to total annihilation. It also means there could be inbreeding problems in the future, so an epidemic of a fatal disease could wipe out the entire population. Their future may not be as rosy as was first thought.

Below and Overleaf
Many species of whales, including southern right whales (below) and humpback whales (overleaf) are often seen leaping from the water, a behaviour known as breaching. Why they breach is unclear. It's thought it could be a form of communication.

Partners until the end of time

Subantarctic islands are where huge numbers of the region's seabirds and seals gather during the breeding season, and Bird Island off the coast of South Georgia is the temporary home for several species of albatross, including the grey-headed albatross.

Most parent birds make a huge investment to raise their chick, so it pays to find the right partner and stick with them for the rest of their life. This is what many albatrosses do, their courtship ceremonies tending to be quite elaborate, first testing a prospective mate and then re-testing at the start of every breeding season. The grey-headed albatrosses display on the ground. Both male and female spread their wings, wave their heads and point skywards, touch flanks and rattle their bills, while making loud braying calls, a behaviour known as 'sky pointing'.

Both albatross parents feed and nurture their young, and grey-headed albatross parents have their work cut out. They breed every two years, nesting among the tussock grass on the island's steep slopes. They feed from the

stormy South Atlantic and Southern oceans, where grey-headed albatrosses go further south than the other albatross species – travelling up to 13,000 km on a single feeding trip, and covering 1,000 kilometres in a day. One bird, monitored by the British Antarctic Survey, circumnavigated the Antarctic mainland in just 46 days. There is food aplenty for the moment, but commercial long-line fishing boats are snagging ever more albatrosses and the weather, due to global warming, is becoming all the more hostile, giving rise to an unexpected danger.

The single, large, fluffy chick sits in a high-sided, mud pillar nest, resembling an elephant's foot, so it is raised above the cold, wet ground. Here, it is relatively safe, but should it fall out or be blown out and cannot get back in, it is doomed – not initially because of predators, but because of the behaviour of its parents. They will not feed a chick unless it is in the nest, and increasingly nowadays chicks are being blown out of their nests by strong winds. Giant petrels clear up the casualties.

The Southern Ocean is notorious for its winds. There is no large landmass to slow them down, so intense and frequent cyclonic storms blow eastwards unheeded. They form because of the temperature contrast between the cold continent with its icy apron and the warmer open ocean, and they are the bane of sailors and albatrosses alike. Between latitude 40 South and the Antarctic Circle, the winds have been given names that reflect their ferocity and latitude: 'Roaring Forties', 'Furious Fifties' and 'Shrieking Sixties'. Between them, they have the strongest annual average wind speeds found anywhere on Earth, and they drive the largest annual average wave heights. The worrying thing is that data gathered during the past four decades shows that surface wind speeds are increasing significantly. They are also blowing stronger than at any time during the past 1,000 years. The result is that more albatross chicks are being blown out of their nests; and the species can't afford to lose a single chick. Bird Island may have the world's largest population of these birds for now, but numbers are dropping at a rate of five per cent a year, prompting IUCN to list the grey-headed albatross as 'Endangered'.

Below
The grey-headed albatross is a fast flyer. When assisted by tailwinds, its average ground speed on foraging trips has been estimated to be 78.9 mph, and it can keep this up for eight hours or more.

Opposite
The grey-headed albatross nest is a tall cone of mud, lined with grasses. Over the years, moss often covers the outside.

Overleaf
Nests are generally on steep slopes or cliffs amongst tussock grass on remote sub-Antarctic islands, and they can be very exposed to the wind and weather.

Behind the Scenes
South Georgia odyssey

Strong winds and high seas greeted the production team as they headed for South Georgia from the Falkland Islands. Among those on board a sailboat specially equipped for exploring the Southern Ocean were producer Fredi Devas and wildlife film cameraman Mark McEwen. For Mark, it was a childhood dream come true. At last, he could follow in the footsteps of his childhood hero, the British explorer Ernest Shackleton.

'When Fredi asked me to do the shoot, I jumped at it. What I didn't realise when I said "yes" is that the journey down there is across one of the most notorious and terrifyingly rough parts of the sea – the Drake Passage. When I got on board, I noticed that everything was bolted down firmly or tied very tight.'

And, true to form, the seven-day crossing was extremely rough. Very soon into the journey, Mark's plan to relax and read while sailing across the deep blue sea was not about to become a reality.

'We were thrown about for the next week; with the sound and smell of the film crew, bedridden and throwing up continuously. It felt like being in a washing machine with a food-recycling bin! The waves were huge, so you felt so helpless in a small yacht. The safety brief was along the lines of "don't fall overboard, because we won't be able to save you in these waters", and we also had to watch out for icebergs.'

When they arrived safely on South Georgia, however, all that discomfort was quickly forgotten. Mark found himself in a place with breathtaking scenery that was packed with wildlife.

'St Andrews Bay is this vast amphitheatre surrounded by towering mountains and glaciers, and there was snow everywhere. When we anchored, the sound that greeted us was overwhelming – the deep bass tones of the male elephant seals mixed with the braying of thousands of king penguins.'

One reason for the huge concentration of animals is that South Georgia sits close to the Antarctic Convergence, where cold Antarctic water meets relatively warm water from the north. The cold water sinks and the warm water rises, creating an area of mixing and upwelling that is extremely productive, so it attracts enormous numbers of marine mammals and seabirds.

The seals rear their young mainly during spring and summer, but the king penguins are busy for large parts of the year because the species has an odd life cycle. On South Georgia, from laying to fledging can take between 13 and 16 months, so, with chicks from the early breeders hatching about two months before the late breeders, there are birds at different parts of their cycle during the entire breeding season, and, as Fredi discovered, they are intensely curious.

'I was sitting down, searching for a piece of kit, and looked up to find a group of king penguins surrounding me. They waddled up and began to peer

Below
St Andrews Bay on the north coast of South Georgia is the temporary home to large numbers of breeding and moulting southern elephant seals and king penguins. Other breeders include Antarctic fur seals, light-mantled sooty albatrosses, snowy sheathbills, brown skuas and Antarctic terns.

into my bags and pecked at the cameras. One started to nibble my glove. I think they were trying to work out what we were!'

For many penguins, the breeding cycle starts when adults return to the colony for the prenuptial moult. Fredi noticed that not all the birds behaved in the same way.

'They come to land to shed their old feathers and grow new ones. It's a big physiological change, and they can seem a little grumpy. Many of them were in huddles on the beach, where the sea breeze blew away their feathers. Others chose to walk inland for about two kilometres and up onto the glacier. It's a long waddle away from the beach, and these penguins were more widely spaced out, but why do they go to such lengths to get away from the crowd? Are they simply less sociable or is something else going on?'

Whatever the reason, those unsuccessful the previous year tend to arrive first. It gives them a head start. When they have found a mate, they tend to be monogamous during the current breeding cycle. Only 29 per cent of the breeding females retain the same mate from one cycle to the next.

A single egg is incubated on the parents' feet. They work in shifts. Males take the first shift, but if the female is back late from foraging and the male has insufficient fat reserves, he might abandon the egg. Hatching can take up to three days to complete, as the eggshell is so thick. The newly hatched chick, though, has relatively thin down feathers and, at first, it is totally dependent on its parents for food and warmth. During this 'guard' phase parents alternate between balancing the chick on their feet and keeping it warm under a pouch, much like emperor penguins do, and fishing out at sea. As the chick grows, it is better able to defend itself against the predators that plague the colony, such as giant petrels and skuas, so both parents can then go fishing.

Above left
Wildlife cameraman Rolf Steinmann focused on the king penguin colony with close to 500,000 birds.

Above
The fine dark sand beach at St Andrews Bay is said to have the greatest mammalian and avian biomass of any beach in the world.

Opposite
King penguin chicks have a thick overcoat of brown down feathers to help keep them warm. Parents return to feed them every 7–14 days, although for late breeders this can stretch to four–six weeks during winter. Chicks lose up to 50 per cent of bodyweight between feeds, although one individual was known to wait for five months and it still survived!

To keep warm, the young penguins wear an overcoat of thick brown down. It gives them good insulation against the cold, but, just to be sure, before they head out to sea, the parents encourage their offspring to bunch together with other youngsters. Being in a huddle keeps the whole group warm, and they are safer in a crowd. On leaving their chicks, the parents have a long walk to reach the shore, and, when they hit the beach, they have to negotiate a wall of blubber.

The world's largest species of seal – the southern elephant seal – breeds on South Georgia's beaches. Over half of the world's population comes here, and they are big… very big. Bulls can be up to 6.85 metres long and weigh up to five tonnes, more than five times the weight of a small car, the vital statistics of a record-breaking bull that was shot here in 1913. Females are about 40 per cent smaller, and there are so many of them at St Andrews Bay, that the beach has one of the highest biomasses of any beach in the world. Space, therefore, is at a premium.

The mature bulls arrive first, and they fight for the right to occupy a section of the shore where the females will eventually move in. They rear up and, using large canines as weapons, slash at the thick blubber around their opponent's neck.

It was Mark's task to film these vicious fights, and he quickly realised just how big they really are. They towered over him. Using a hand-held stabilised camera system to reduce shake, shudder and wobble, he was still able to get in very close, right in the thick of the action.

'They would charge at each other surprisingly quickly for such an enormous animal. I had to get out of the way mid-shot, and the shudder when they hit could be felt through the ground. It reminded me of that scene in *Jurassic Park* when the footsteps of T. rex cause the water to ripple.

'Fortunately, I had Fredi watching my back. He would pull me back when things got dicey, as you get so immersed in the moment that you don't see what else is happening around you. Frequently, I'd be filming one pair fighting, unaware that I was about to walk into others about to fight.'

The contests can be bloody, the teeth causing deep gashes in the skin, but the confrontations rarely end with the death of one of the bulls. The weaker male usually backs down. The winner waits to welcome his harem, when each dominant male or 'beachmaster' gathers a bevy of females about him. They give birth almost immediately, mother and pup exchanging calls to help establish a bond. The pups yap and their mothers moan, and the youngsters became unexpected obstacles while Mark was filming.

'They would often crawl up the beach and spend the day resting under my tripod, and a metre-long, 80-kilogram seal becomes quite a trip hazard when you're running out of the way of charging bulls... but they were incredibly cute, sometimes imitating the grown-ups and play-fighting on the beach.'

The beachmaster cannot mate immediately. He has to wait for his females to become fertile again, usually a few days after the pups are born. At this time, he cannot feed, maybe for up to three months. He dare not leave the beach, so he must rely on his fat reserves. He loses up to 15 kilograms a

Above
When walking to and from the sea, the king penguins have to run the gauntlet of elephant seals, and be particularly wary of the restless bulls.

Overleaf
Two large and equal-sized bull elephant seals square up to each other. After an exchange of belching snorts, they will come to blows.

Above
Wildlife cameraman
Mark McEwen uses
a gyro-stabilised film
system to get in close
to the action as two
bulls are about to fight.

day, for he must keep active, seeing off other males who will try continually to usurp him or even try to mate with his females while he's not looking. Even though he is not the father of the present litter, he wants to ensure he is the father of the next. He threatens them with a guttural blast of sound and rises up to his full combat height. This is usually enough to put down an insurrection, but just occasionally two bulls are equally matched, and another fight ensues. Any pups in the vicinity are suddenly vulnerable – after all, he is not their father – and while the film crew were on the beach in St Andrews Bay several pups were crushed.

The king penguin parents, meanwhile, have been fishing far out at sea. They will travel about 30 kilometres from the colony during a single excursion, diving down to about 300 metres in pursuit of squid and fish, particularly lanternfish. If they avoid leopard seals and orcas, which sometimes patrol inshore, and make it through the elephant seals without disturbing a beachmaster, they then have the difficult task of finding their offspring amongst the hundreds of thousands of birds in the colony.

Key to the chick's survival at this stage in its life is that it does not give in to an urge to explore, and not move too far from where its parents left it. Parents find their chick not by sight or smell but by the calls it makes. However, an individual's call can get lost in the cacophony. Add to that the roar of the fierce katabatic winds that sweep down from the mountains and glaciers and the pounding of the surf, and the chances of a youngster getting lost are high.

They must stay close to where they were last fed, often within a crèche, if they are to have any chance of being found.

When the youngsters moult their overcoat and reveal their adult plumage, they are abandoned. It is their turn to run the gauntlet and make for the open sea. And, it seems the species, along with the elephant and fur seals living on South Georgia, is doing well. It was clear to Fredi that at least some of South Georgia's wildlife is bucking the global trend of animals heading for extinction.

'I had been to St Andrews Bay nine years earlier, and there are very few places in the world where you return to after nearly a decade and find that the wildlife is doing even better. To me, it felt like there were more king penguins and elephant seals, and we saw whales on the journey down. Since the ban on hunting whales, seals and penguins came into force, and since rats and reindeer have been removed from the island, native species are most definitely benefitting.'

However, climate change and global warming is evident even in this wildlife paradise. The Cook Glacier, which flows into St Andrews Bay, is a casualty.

'The glacier in the bay has retreated significantly. So, although things for some species may be looking better right now, it is clear that climate change is starting to have a big impact on the wildlife of the Antarctic.

'St Andrews Bay is such an unbelievably special place, that it reminds you of the value of these wildernesses, and we must try to keep these in mind when we are making decisions on the other side of the world that will have an impact on their future.'

Below
A sub-adult bull elephant seal, surrounded by king penguin adults and chicks, waits between 'practice bouts'. It could be another three or four years before he is physically capable of challenging a beachmaster.

Asia

Producer: Emma Napper

Left
During the summers
of 2004 and 2005, the
surface of the Lut Desert
of Iran was baked by
the sun, so satellite
sensors could reveal
that this is the hottest
and driest known place
on the planet.

Introduction

Covering about 30 per cent of the world's land surface, Asia is the largest of the continents. From east to west, it extends across eleven time zones and, from north to south, it stretches from the Arctic to the tropics. It is a continent of vast open spaces and of extreme weather and climate.

The world's coldest permanently inhabited town is Oymyakon, in northern Russia, which has an air temperature that once dropped to minus 67.8°C, while in the hot deserts in southwest Asia satellites have measured the world's highest ground temperature: 70.7°C in the Lut Desert of Iran. The wettest place on Earth is Mawsynram in tropical India, with an average annual rainfall of 11,871 millimetres, and the deepest snow accumulates regularly on the Japanese Alps of Honshu Island, Japan, where a world record 11.82 metres fell on the slopes of Mount Ibuki on 14 February 1927.

The world's highest mountains are the Himalayas and their northwestern neighbours, the Hindu Kush and Karakorum. Together, they form a barrier to the movement of wildlife north and south – the tundra, taiga and steppe in the north are separated physically from the mainly tropical and subtropical deserts, forests and islands in the south. The only animals to make it over the top are migrating birds, including bar-headed geese, which fly close to 6,000 metres above sea level, where the air is thin.

Asia has some of the densest concentrations of people on the planet, too, so the natural fauna and flora close to major conurbations has had to adapt to living alongside many people with all the dangers that entails. Habitat destruction and poaching have brought many animals, including several subspecies of the rhinoceros and tiger, to the brink of extinction. Even so, in some places, plants and animals thrive practically undisturbed because the continent is so big and some areas so remote.

Geyser bears

On its eastern margins, the Asian continent appears to be alive. Some sections of the east coast are part of the Ring of Fire, the volcanically active region that surrounds the Pacific Ocean. One place that you see and feel this incredible power of nature is on the Kamchatka Peninsula in the far east of Russia. Here, the eight-kilometre-long Valley of the Geysers in the Kronotsky Nature Reserve is the second-largest geyser field in the world. It has more than 40 geysers and countless pulsating hot springs, gushing vents and mud pots. Most are on one bank of the Geysernaya river, its waters warmed by geothermal water running off Kikhpinych, a young stratovolcano with a sinister reputation. At its base is the Valley of Death, a place where an accumulation of noxious gases kills almost all living things that venture in.

The pattern of deaths seems to coincide with the season. First small birds are killed in snow hollows, followed by foxes attracted to the dead bodies. They, in turn, succumb to the poisonous gases. Bodies do not decay readily as bacteria are also affected, so the deadly procession sees the arrival of other opportunistic scavengers, such as wolverines, bears, crows and eagles, and they die too. Most deaths are on still, sunless, snowless days between May and October. Downstream, the Valley of the Geysers is less troublesome, but at one point it was almost lost for ever.

In 2007, a mudslide, caused by the collapse of the land around a waterfall, almost blotted out the valley and its thermal features, and created a turquoise thermal lake upstream. However, much of it has survived, including Velikan,

the giant geyser, which produces a jet of water up to 40 metres high every two to three hours. In early spring, there's every chance you might encounter brown bears right beside one of those eruptions, almost hidden by the steam. Cameraman John Shier was there to meet them.

'It's an immense landscape, and you get an overwhelming sense of just how desolate and barren it is. But, when you see the steam, you have a shock of colour in this monochrome world.'

It is warmer here than in the surrounding countryside. Everywhere else around is still covered by ice and snow, but here it's green. When you lie down, the ground is warm, the temperature just 500 metres down being up to 250°C, but when you stand up the air is freezing. Even so, brown bears emerge from their six-month hibernation a little earlier here than elsewhere in Siberia, giving them a significant head start to fatten up and regain their strength ahead of their rivals. Mothers and cubs lie prostrate on the ground, bathing in the heat, and they even seem to relish the steam bath they receive from the geysers.

'Big males and single females claim the best patches of grass,' John noticed. 'Smaller males keep an eye out and grab a bite where they can. The bigger males will chase and attack them, but those that must be most wary

Above and Overleaf
The bears roam in an area of geothermal activity that is second only to Yellowstone in extent and intensity.

are mothers with cubs. They tend to stick to the high ridges where they find patches of grass and roots to dig out. Above all they must watch out for males for they will kill cubs. Mothers must always be weighing risk and reward, and being too cautious can be as deadly as being too bold.'

There are about 800 brown bears in Kronotsky, the largest protected population of brown bears in the whole of Eurasia. Some are monsters, almost the size of Kodiak brown bears, the biggest bears in the world. Even so, they are considered to be harmless to people, although there was one occasion in July 2008 when 30 bears besieged a platinum mining compound, killing two guards and preventing the inhabitants from leaving their homes for several days.

Some of the bears also have addiction problems. Discarded barrels of aviation fuel litter the ground where helicopters land and refuel, and the bears have learned that if they pierce the cans and sniff the fumes, they get a 'high'. When intoxicated, they lie on their back in what Russian nature photographer Igor Shpilenok described as the 'nirvana position'. They even stalk helicopters in order to sniff any fuel that drips down when the aircraft is refuelled. Kamchatka has delinquent bears!

Mountain monkeys

The rugged mountain forest in China's Shennongjia National Park is not the obvious place to find monkeys. Winter temperatures can drop to minus 8.3°C, and snow cover lasts for four to five months of the year, yet one of the world's rarest monkeys survives up here. Feeding on lichens, and whatever else the forest has on seasonal offer, the golden snub-nose monkey lives at altitudes up to 3,400 metres, where it withstands the coldest average winter temperatures of any non-human primate.

The monkeys wander about in family groups, each with a male and several females and their young, and they are enormously social animals. In summer many groups band together, so there might be up to 200 individuals travelling and feeding in a single enormous troop. They spend about 97 per cent of their lives in the trees, travelling and foraging during the morning and afternoon, while stopping for a siesta at midday. They chatter amongst themselves, producing a great variety of sounds, but do so without any facial movements, like ventriloquists.

When on the move, males are in the vanguard and rearguard, for there are many predators waiting and ready to take them. Goshawks and golden eagles swoop in to snatch youngsters from the trees. Asiatic golden cats and weasels are up there hunting too, while wolves, dholes, and leopards pounce on any monkey caught on the ground. The biggest threat to the monkeys,

Above
An infant golden snub-nosed monkey has light grey-brown fur at first, which gradually darkens as it gets older.

Overleaf
A group of snub-nosed monkeys all huddle together for warmth in mountains where the winter temperature can drop to minus 8.3°C.

however, is people. Logging – legal or illegal – destroys their natural habitat, and just the sound of chainsaws can see them moving elsewhere, disturbing their ecology. Snub-nosed monkeys are listed as 'endangered' by IUCN.

In winter, competition between family groups for limited food supplies means that any harmony that had been there in the summer breaks down during the day, although they are still found huddling together for warmth at night. Males, with very obvious large canines, fight for access to the best resources. The contests are accompanied by a lot of bravado and shouting, when the snow flies about and families egg on their champion, but there is generally little physical contact between the combatants. Missing tails, however, indicate that they do come to blows occasionally, and these rare fights must be extremely violent.

These warriors, however, no matter how brave, seem to dislike getting their hands cold. They walk upright in the snow, cartwheeling their arms, and the really strange thing is these bipedal monkeys look remarkably like the popular image of abominable snowmen. Are they the real yetis?

Below
An infant spends a large amount of its time playing close to its mother or feeding on her milk. Mother might have helpers to assist in the care of her young.

Along came a spider

In the southwest corner of Asia, the deserts of Iran are among the hottest and driest places on Earth, but there are a few oases where plants can grow and animals can live. There are resident birds that nest amongst the rocks, and migrants that are just passing through. Many of the smaller birds catch spiders and insects, but sometimes their prey is not what it seems.

A bird might fly down to take a look at a spider wriggling on the hot rocks and be just about to grab it when a snake's mouth unexpectedly appears and grabs the unfortunate bird, swallowing it whole. The victim has been duped by none other than the spider-tailed horned viper, which is aptly named. Its tail ends in a bulb bordered by appendage-like scales, which together have a striking resemblance to a spider. The snake also moves its tail so the bogus spider 'comes alive'; however, it does not fool all of the birds all of the time.

Resident birds, probably the descendants of birds that were fast enough to have survived attacks, know not to approach the mock spider. They are rarely caught, and live in relative harmony with the snakes. The victims are generally migrants and, because of this, the snake itself has altered its pattern of feeding. It is so attuned to events in its environment that it only catches birds for about three weeks during the birds' northward migration and for another three weeks when they are heading south. It is also careful where to lie in wait. The incoming birds tend to fly at particular heights so the snake sits in wait on ledges at different heights, depending on which birds are passing through. Using mimicry, camouflage, and knowing the behaviour of prey are ways to survive in such a hostile environment.

Below and Opposite
While the spider-tailed horned viper's body colour and pattern help it blend in with its background, the spider-like scales at the end of its tail are designed to be very obvious to passing birds.

King of the castle

On a dry grassland plateau in the Western Ghats of southwest India, a male fan-throated lizard, known also by its genus name *Sitana*, is trying to be conspicuous. He wants to attract a mate. He has a fan-shaped, iridescent blue dewlap under his chin, which he flashes at any prospective partner, but for the moment his colourful display fails to impress any of them. As a female approaches, he climbs onto a small rock and tries again. The female races off on her hind legs at high speed, bipedal movement being a frequently used mode of locomotion for this species. The male's fan, and no doubt his ego, deflates. He finds a slightly taller rock. The same thing happens. The female is unimpressed. He has to find a better rock. When he spots just the right one, he sees that another male is occupying the best position, and he has quite a following. The only way he is going to be in with a chance is to fight.

He approaches the other male, raising a crest on his back and another on the back of his head. He moves with a jerky, stiff-legged gait, showing he means business. When the other male does not back down, they are suddenly at each other's throats. Open-mouthed, they grab each other, shaking their opponent's body violently. These fights can end in the death of one of the contestants, and many die during the breeding season, but this time the challenger wins without the shedding of blood. His opponent beats a hasty retreat, upright and on two legs, of course. Now our young beau can really impress the opposite sex. He is king of the castle, and he'll remain that way until a dirty rascal can topple the king.

Above and Opposite
The male fan-throated lizard has an extendable dewlap with iridescent blue, black and orange patches and yellow stripes. The orange patches are attractive to females, while the blue and black colours signal to other males to keep their distance. The male is highly territorial, and might make a hasty retreat after a failed encounter with another male by running away on its back legs.

Person of the forest

In Southeast Asia, the monsoon rains bring life to the lush tropical rainforests of Kalimantan, the Indonesian part of Borneo. The island boasts some of the tallest tropical rainforest trees in the world, up to 100 metres high, and they are home to a very special animal, the very rare Borneo orangutan. The name comes from the Malay and Indonesian *orang*, meaning 'person', and *hutan*, meaning 'forest', and it refers to one of the most intelligent of the primates, with a great capacity for learning. In the forests here, the filmmakers followed a 40-year-old mother, called Bibi, with a two-year-old youngster, known as Bayas. Both have been observed by scientists for a number of years. They watched as the mother taught her offspring about the vast and complex place in which they live. Its life, understandably, depends on it, for it is a place where just moving around is not as straightforward as one might think.

The forest canopy can be a death trap, especially when clambering between trees. For this, a youngster needs its mother's help, and one of the fundamental things she can do to ensure her offspring's safety is to build bridges. The gaps between trees are no problem for an adult orangutan – the most arboreal of the great apes – but they can be insurmountable for youngsters. Mother has a simple solution. She bends branches to bridge the gap, so the two of them can travel through the trees without mishap.

The baby must also learn what to eat and how to eat it, and it does this by watching and copying its mother's behaviour. The adult takes a rotting branch and sucks out the termites inside, a tasty snack. Copying its mother, the baby takes a smaller branch, more a twig than a branch, but really it is clueless about what it should do. It needs to watch and learn a whole lot more.

The youngster must also be in tune with the sounds of the forest. Animals pitch their calls for optimum transmission through the trees, so sounds can carry over great distances, and the long call of

Right and Overleaf
While riding with mother is the safest way to travel from branch to branch, eventually the infant orangutan must learn how to tackle the tangle of branches itself, but always with mother keeping a close eye on her offspring.

the male orangutan is one for the baby to learn. He's a powerful animal, recognised by his large cheek pads, and his call is designed to attract females while scaring off rivals. A young orangutan, whether it is male or female, must be able to recognise the sound, for its life in the future will be influenced by it.

The young orangutan's last lesson of the day is to build a night nest, which is an elaborate platform set amongst the branches. The young orangutan must be particularly attentive. When it grows up, the nest will need to support the weight of an adult – 37 kilograms for a female and 75 kilograms for a mature male. If it gets it wrong, it will end up in a pile on the forest floor.

Not many do get it wrong. Research has revealed that orangutan mothers are probably the best in the world. About 91 per cent of all offspring survive until they are weaned, which occurs at about eight years old, and, of these 94 per cent of weaned females make it to adulthood and have babies themselves. However, the figures for male offspring are not known because once they have left their mother they are hard to track. This is a better survival rate than for any other species of great ape, and even for some human populations.

Below
An infant orangutan learns what is good to eat and what is not, partly by example when watching mother, and partly by testing potential foods for itself.

One reason is that the forests of Southeast Asia are less productive than one imagines. The availability of fruit is sporadic, so to find sufficient food without having to travel too far, adult orangutans are solitary. This reduces competition and aggression, which inevitably occurs in groups, and helps isolate them from infectious diseases. Living in the trees also helps keep offspring safe from predators.

The downside is that orangutan mothers can only rear one infant every seven years, so a population is very slow to recover. Borneo's orangutans are 'critically endangered', according to IUCN's Red List, with the country having lost more than half of its orangutans during the past 16 years, due mainly to deforestation. There are an estimated 10,000 remaining, which sounds healthy, but they are actually in 64 small populations isolated from each other by roads, rivers and plantations. Of those 64, only 38 have more than 100 animals, so are viable as breeding populations. It seems surviving mothers are going to need all the help they can get.

Opposite
The orangutan has a relatively long childhood. It will be with mother until it is at least eight years old.

Dinner bell for giants

To the east of Borneo, located in the waters bordered by Indonesia, Malaysia, Papua New Guinea, the Philippines, Solomon Islands and Timor-Leste, is the spectacular 'Coral Triangle'. It is a hotspot of biodiversity, one of the most important centres for tropical marine life in the world. Over three-quarters of all the known species of coral and over half of all Indo-Pacific species of reef fishes are found here.

Located within the triangle, on the north coast of Papua is Cenderawasih Bay. It has become famous for its gigantic visitors and a welcome conservation success story. The main characters are whale sharks, the world's largest living fish, and they come here for a free food handout. Watching was director Lucy Wells, and it was the sheer size of these creatures that took her aback.

'When the opportunity arose to swim with them, I was able to appreciate their size even more. Their tail fin was as big as me – nothing prepares you for that!'

Although enormous, up to ten metres long on average, whale sharks are gentle filter feeders, consuming plankton, krill and small fish. While places like Ningaloo Reef in Western Australia, Mexico's Yucatan Peninsula and the Belize Barrier Reef see them for two or three months in the year, Cenderawasih has them all year round, and they gather around the semi-mobile fishing platforms used by local fishermen, known as *bagans*.

Below
Whale sharks may be huge, but they are harmless filter feeders so being in the water with them is not dangerous, though divers do have to watch out for a sweep of their powerful tail.

Opposite
Whale sharks have several methods of feeding: one is to hang vertically, open their mouth, and let water and small fish flow in, like water going down the plughole.

Overleaf
Quite a crowd gathers when the sharks are freeloading.

Above
Local fishermen hand feed the sharks. They believe it brings them good luck.

Opposite
Where whale sharks go when they leave the bay is a mystery; in fact, where they go and what they do for most of their lives is generally unknown.

The current crop of fishermen has been fishing in Kwatisore Bay, in the southwest of Cenderawasih, for about 25 years, and the whale sharks were in the bay when they first arrived and erected their bagans. Now there are 23 platforms, from which huge nets are lowered to a depth of about 18 metres at dusk. Floodlights on the surface attract millions of anchovy-like baitfish, known as *ikan puri*, which are caught in the nets. The following morning, the nets are winched up and unloaded, but not all the fish are removed. Some of the anchovies are left in and the net lowered just over the side, for these fishermen have changed from killing whale sharks to feeding them.

The whale sharks have learned that they can suck the fish out through holes in the mesh, and a dozen or more sharks might be found around a single bagan doing just that. They have even started to beg, rewarded by a bucket of fish, for the fishermen believe it brings good luck. Some of the anchovies are used to catch the tuna, bonito and other game fish that are attracted to the feeding frenzy, so the whale sharks appear to bring good fortune.

Some sharks have been tagged, and this has revealed that many are not residents, although some return to the bay each year. The danger now is that shark-fin merchants arrive and get rogue fishermen to kill off the sharks for their fins, which go to make shark-fin soup.

So far, they have been kept out of the bay, for it is protected as a national park, the biggest in Indonesia. In the distant past, it was an isolated sea in which its own species of corals, reef fish and other marine organisms evolved. It has led some biologists to refer to the Bay as the 'Galápagos of the East', and veteran underwater explorer Valerie Taylor has called it 'the eighth wonder of the world'.

Behind the Scenes
A very rare rhino

On the island of Sumatra, a strange sound carries through the forest. At first, it is remarkably like the song of the humpback whale, but it is not. It is a female Sumatran rhino, one of the most endangered mammals on the planet, and she is looking for a mate. Ambling through the forest, she comes up against a fence. It is a protective fence designed to keep her safe from poachers. With few rhinos left in the wild in a sustainable population, she is vital to her species. Her location is a secret, but producer Emma Napper and her film crew were given access.

'Sumatran rhinos are so rare – fewer than 80 individuals living in the wild – that we weren't even sure of the exact location. Hidden inside a national park, the area is off limits to all but a few people.

'I wanted to hear them sing, an adaptation to the thick forest in which they live, and as soon as we drove to the secret area, I could hear it – beautiful, melodic and mournful. We eventually got a glimpse of her – Ratu. She's been protected here since her forest was destroyed, and she's tiny! I'm only 5ft 2in, and she's not as tall as me.'

In order not to disturb this valuable female, the film crew were not allowed into her large enclosure; they had to erect a cable dolly, with a camera hanging from zip wires strung between trees – low enough to see her through the trees, but high enough that she didn't poke it. In this way they could 'fly' over the enclosure and obtain close shots of the female rhino without disturbing her.

'It meant putting two cable operators up in a tree close to the enclosure, but they could only come down if we were sure Ratu was having a nap.'

The Sumatran rhino is more active at dawn and dusk, spending the best part of the day in a mud wallow, essential to prevent its skin cracking and to rid itself of external parasites. The mud also helps in temperature regulation. This species seems particularly sensitive to high temperatures so, during the warmer rainy season, it will seek out higher ground and cooler temperatures.

These rhinos communicate with others of their kind by scraping the ground with their feet, leaving dung piles and twisting saplings. They make three distinct sounds: short 'eeps', the humpback-whale-like sound, and the very loud whistle-blow. The whistle-blow is so loud that it can be heard over great distances in the forest. The only hope now is that there are enough Sumatran rhinos in the wild to answer those calls.

'It was a privilege to see such a rare animal, and sad. It's hard to imagine that there are fewer Sumatran rhinos in the world than there are people in my street. There's a real chance that these animals will be gone for ever, during my lifetime.'

Below and Opposite
The female rhino the team filmed was in a large enclosure to keep her safe from poachers. The film crew were not allowed to enter, for fear of disturbing her, so they had to rig up a complex camera system strung from wires between the trees above her enclosure.

Africa

Producer: Giles Badger

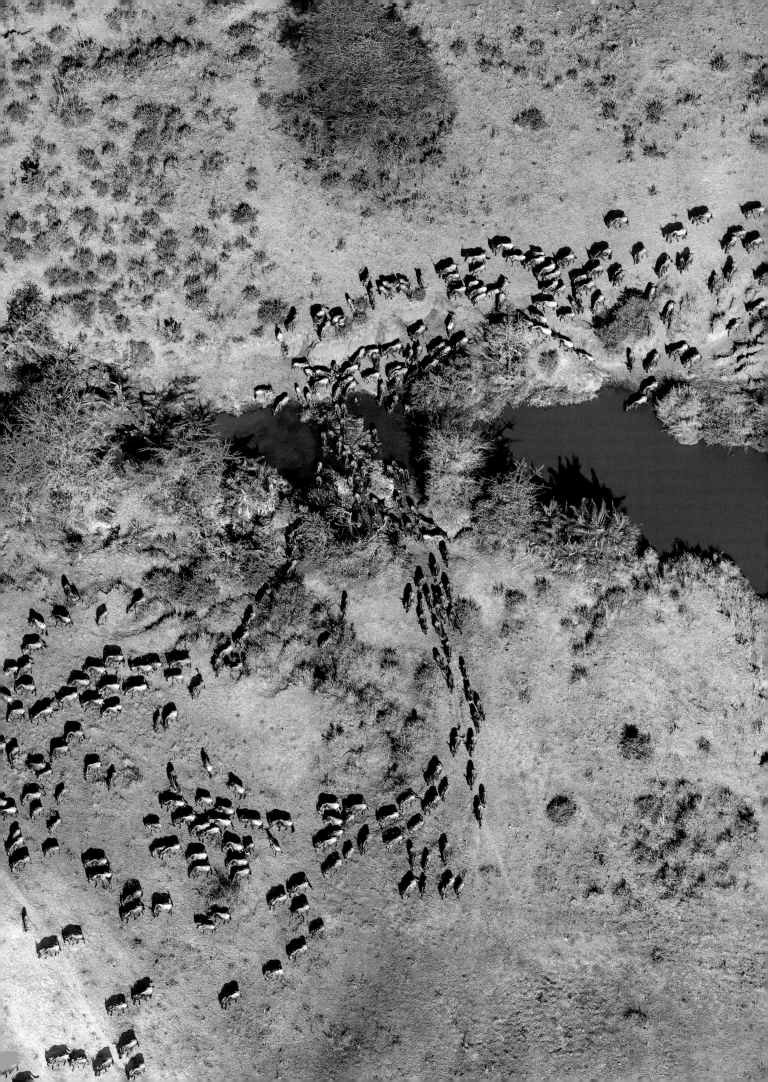

Left
A herd of wildebeest at a river crossing in the Masai Mara, Kenya. They're just a small part of one of the greatest migrations of land animals on the planet.

Introduction

Africa accounts for about 20.2 per cent of the Earth's landmass, making it the second-largest continent, but for wildlife extensive areas are hot and inhospitable. The scorching Sahara, the world's largest hot desert, and the less extreme but still dry and windy steppe of the Sahel dominate Africa's northern half, with the Namib and Kalahari deserts in the southwest and the Danakil on the Horn of Africa in the east. Many parts of Africa are vast savannahs that stretch out as far as the eye can see, like those across East and Southern Africa, while straddling the Equator is the basin of the Congo River and the second-largest tropical rainforest on Earth.

On these lands the world's largest populations of free-ranging megafauna on non-private lands live. Here, elephants, rhinoceroses, hippopotamuses, antelope, big cats, monkeys and great apes are still to be encountered, although populations are falling due to illegal hunting, the spread of diseases, changes in climate, and habitat destruction. It is estimated, for example, that Africa is losing its forests at twice the rate of the rest of the world.

Without a doubt, many parts of Africa are difficult places to live, and for pioneers and survivors, key issues are to know where to find water... and to avoid humans, which is not so easy, as with more than 1.2 billion people, Africa is the second most populated continent on Earth.

Africa is generally accepted as the birthplace of *Homo sapiens*, our species splitting away from our nearest hominin relatives, the Neanderthals, maybe as early as 800,000 years ago. It is thought several waves of human migrants left the continent and went on to colonise the entire world.

Ghost towns and seal nurseries

The Namib rivals the Atacama for the title 'oldest desert in the world' and, with an annual average rainfall of just 5 millimetres in the west and little more than 85 millimetres in the east, it is the only true desert in Southern Africa. Surface water is almost nonexistent. Most rivers flow underground. The close proximity of the Atlantic Ocean, however, means that early morning mists and fogs bring moisture, so small plants can survive here, and many of the Namibia's animals are adapted to living in such dry conditions.

On the Namib's Skeleton Coast, there are not only the skeletons of shipwrecks, but also the skeletal remains of buildings. These ghost towns, were once home to diamond miners, but after the Second World War, fewer diamonds were found and everybody left, heading to better diamond fields elsewhere. The towns were totally abandoned by the 1950s, and, since then, the desert has reclaimed them.

Brown hyenas have replaced miners and their families. They hide amongst the sand-filled rooms to escape the heat of the day, emerging when it is cooler to search for food. This species has a different appearance to the spotted hyenas more usually seen in wildlife films. It has long shaggy fur, pointy ears and erectile hairs on the neck and back that can be raised when it wants to appear bigger and more threatening. Like its spotted relatives, it has powerful

Below
Brown hyenas have replaced diamond miners in some of the ghost towns of the Namib. Normally found in rocky areas that provide shade, the hyenas shelter in derelict buildings from the midday sun.

Opposite
When the diamonds dried up the miners moved elsewhere, and now the Namib Desert is reclaiming the land. Sand fills buildings, and the only sound is wind blowing through the ruins and the roar of the surf.

jaws, but it is a poor hunter. The brown hyena relies mainly on prey killed by other predators, such as cheetahs, and gets all the water it needs from the carcasses, but in the Baker's Bay area on the Namib coast, hyenas have become accomplished predators, for, along the shore, they have a convenient meat store. A large population of the world's Cape fur seals haul out here, and they have pups. The hyenas hunt the vulnerable babies and scavenge the bodies of any dead seals. It just goes to show how there is food in the most inhospitable places, and sometimes plenty of it… you just need to know where to look.

Below
The hyena may have travelled tens of kilometres across the desert to reach the Cape fur seal colonies at the coast.

Opposite
These hyenas are generally scavengers. They have extremely powerful jaws and, although poor hunters, they can tackle a baby seal. It's a seasonal glut, when seals are giving birth, and only accounts for about 3 per cent of their food.

Athletic elephants

At Mana Pools in Zimbabwe, there is plenty of water, yet the land is parched and its wildlife thirsty. The mighty Zambezi is on the doorstep but, for many animals, the river is just too dangerous to approach. Nile crocodiles lurk here, some up to five metres long, and, during the dry season, each of the pools at Mana is likely to have a giant reptilian occupant. It means that, during periods of drought, food and water is in short supply for many animals. This is when terrestrial giants step in to lend a hand… or a trunk!

The elephants at Mana Pools represent one of the largest concentrations of African savannah elephants in Zimbabwe, and it is their interest in the apple-ring acacia or ana tree and its apple-ring-shaped pods that help some of the smaller plant eaters in the area. The tree itself has a long taproot, 20 to 40 metres deep, so it is resistant to droughts. It still functions when all others have shut down, because it has an unusual annual cycle.

The ana tree loses most of its leaves during the wet season, and grows them, along with its dangly yellow flowers, at the beginning of the dry season. It is often the only tree with flowers and leaves during droughts, so its nectar is vital for pollinating insects and the birds that catch them, and plant eaters browse its fresh leaves. Its fruiting pods then mature at the end of the dry season, a critical time for herbivores for conditions are getting really tough.

Inevitably, the fruit on the lower branches is quickly eaten up, and the upper parts of the tree look to be out of reach, even to an elephant's long trunk, but some of the pachyderms have found a way to stretch that little bit further. Like elephants at a circus, they stand on their hind legs to reach the higher branches, and in doing so scatter pods all around, providing other hungry beasts with scraps from their table. It puts a great strain on their bodies, and some old bulls have damaged backs by doing it, but goes to show the importance of this method of feeding; and the elephants make the lives of other Mana Pools inhabitants just that little bit easier.

Opposite
This bull African elephant has learned how to reach the highest fruits by standing on its back legs; no mean feet for an animal so heavy.

Friends in need

A lack of water can force animals to find novel ways to survive, and in Tswalu Kalahari Reserve in North Cape, South Africa, there is a character that simply seeks the help of another, totally unrelated, animal.

This privately owned game reserve is trying to re-wild this part of the country, with introduced black rhino and black-maned Kalahari lions, but one of its lesser-known inhabitants is a tiny bird – the ant-eating chat. As its name suggests, it feeds on ants and termites, as well as grasshoppers, beetles, butterfly and moth caterpillars, millipedes and a little fruit. For most of the year it is surrounded by its food, but in midwinter, when it is cool and dry and insect food is hard to find, the bird looks out for pangolins and aardvarks. It might start by shadowing the pangolin, but when it spots an aardvark, which can dig deeper and faster, it switches allegiance.

The aardvark has powerful forelegs and claws with which it tears into termite and ant nests, but is mainly active at night, when the ant-eating chat is tucked up in its roost. However, in winter the aardvark is out and about during the late afternoon, and the bird has noticed its new schedule. When the aardvark has finished digging and eating, the bird drops down to catch the insects the aardvark has disturbed. It is a classic example of commensalism, in which the chat relies on the digging behaviour of the aardvark, and, in doing so, it has discovered that it can take advantage of its new companion at other times too. During the rest of the year, when the aardvark is mainly nocturnal, insect activity around the breach in their nest sometimes continues until the following morning. Even though there is other food readily available, the chats do not turn up their beaks to an occasional ant or termite fix.

Above
The ant-eating chat follows the aardvark and feeds on the ants and termites that are trying to protect their colony.

Opposite
The aardvark has a sensitive long snout, rather like a pig's, with which it sniffs out ants and termites. Using its sharp claws and powerful forelegs, it breaks into the hard termite mound and licks up the insects with a long sticky tongue.

The alliance

The Masai Mara in Kenya is defined very much by the seasonal availability of water. Here, the huge herds of wildebeest, gazelles and zebra follow the rains, and in doing so trek in a great clockwise circuit of the Serengeti and adjoining lands, one of the greatest land migrations in the world. Watching them is a coalition of five male cheetahs – two sets of brothers and an unrelated male. They have found that, by sticking together and hunting much like a lion pride, they not only retain a large territory with access to food and any females passing through, but they can also bring down animals considerably larger than themselves.

Many cheetahs, especially females, hunt alone or with older offspring and target the smaller antelope – gazelle, dik-dik and baby antelope. Young males often form coalitions, but a hunting group of five is extremely rare. By sticking together after leaving their mother, they can tackle some hefty animals, such as fully grown ostriches and wildebeest, and now this coalition has added topi to its repertoire.

Topi are large, fast and powerful antelope. Their horns are ringed and lyre-shaped, their back slopes down from head to rump, and there is a distinct hump at the shoulder. Males can be solitary, but at breeding time they live in small loose herds consisting of the male and several females. At this time, they defend a territory. The male demarcates his patch with dung and urine,

and daubs a secretion from his suborbital glands on vegetation. The species is famous for the way one member of the herd might stand on a termite mound and survey the surrounding countryside, so to have any chance of success the cheetahs must have an element of surprise.

Hunts sometimes start with the brothers gathered at a known landmark, such as a tree, which they sniff and scent mark in turn, and then maybe roll in the grass. Sometimes a herd of inquisitive wildebeest will approach them, stop a short distance away and snort. It's as if they are saying, 'We know you're there, and there are lots of us!' However, they usually lose their nerve at the last minute and turn tail, but they are careful not to run. That could trigger a chase.

A hunt generally starts when one cheetah moves off. The others follow, and then they spread out across the savannah. They amble along at first, looking for a target. Then, at some hidden signal, the lead cheetah will start to run, slowly at first. The others keep pace with him. If he has spotted something amongst a herd of topi, the others will take their lead from him. As the animals start to panic and run in all directions, one topi stands fast. Bad mistake. The cheetahs are on it in seconds. They jump all over the unfortunate beast. One – known as d'Artagnan to local guides – grabs it around the neck and holds on, applying a chokehold in order to throttle it. Others clamber onto its back and grab its rump, attempting to pull the topi to the ground, but it is a tough animal and remains standing for many minutes. It takes all five of the brothers to eventually topple it; and then it's all over.

Bed and breakfast birds

In South Luangwa National Park, hippos attract unusually large flocks of oxpeckers. These unassuming birds – one species with a red bill, the other with yellow – are always in the background in most wildlife films set in Africa, but here they have come to be centre stage, plucking ticks from their host's skin; at least that was what we always thought they were doing. What they are really after is blood. They divest the host of its parasites but also peck at the skin, keeping sores open and the blood flowing. It means we ought to view oxpeckers in a different light. Yes, there is some kind of mutuality going on. The birds hiss when danger threatens, for example, which warns their host, and they pluck off ticks, but they are also cheeky parasites.

Hippos, though, are not necessarily the best hosts. They are big and have thin skin, but they have violent fights in the river and have a tendency to whirl their excrement around. The rewards may be high, but the costs are

Left
Many different species of large mammals play host to red-billed oxpeckers, including warthogs.

Below
Some yellow-billed oxpeckers remain with their giraffe hosts 24/7. At night they roost between the giraffe's hind legs.

too. There are several more convenient mammals around, and what better than a giraffe? The views are good and there are plenty of ticks lurking in ears, noses and nether regions; in fact, it is so good oxpeckers compete for a place on their back, and not just for feeding.

The red-billed oxpeckers roost in trees at night, but camera traps put out by scientists have revealed that yellow-billed oxpeckers roost on their host. This might be because the red-billed ones are not fussy about who they divest of parasites and blood, while the yellow-billed birds prefer giraffes or buffalo. If they hung out in the trees at night, their host might be long gone in the morning, so they stay put. Images also show that the birds settle between the hind legs of the giraffe, where it is probably warmer and also safer from nocturnal predators, so the yellow-billed oxpeckers get a very good 'bed and breakfast service'.

A cuckoo in the mouth

The East Africa Rift Valley is peppered with many lakes. They include some of the oldest, largest and deepest lakes in the world, and they are natural laboratories in which to study how new species evolve. The animal providing the answers is a fish, known as a cichlid, and the production team discovered that several Rift Valley lakes, such as Lake Tanganyika and Lake Malawi, are packed with them. The lakes are huge, more like mini-seas, and sitting on the bottom in the shallows, director Jo Haley could see how the cichlids transform large swathes of the lakebed.

'It felt almost lunar, pockmarked with huge craters and sandcastles. The architects of these mysterious structures were male cichlids, which built them to attract females during courtship. Females would inspect their creations and select the male with the most impressive one, often the largest. Some of the fish were no more than a few centimetres long, yet they built mounds or excavated hollows that were a metre across. From our drone, they looked like crop circles through the clear blue water.'

The underwater mound builders are among more than 1,500 known species of cichlids – 700 living in Lake Malawi alone – and new ones are being discovered every year. One feature that binds them together is parental care. These fish look after their young and many are mouthbrooders. The female

parent does not place her eggs in a nest but hides them in her mouth. The behaviour is probably the result of crowded lakes, where egg predation is a constant threat. What better place to hide the eggs, and later the fry, than in one's own mouth? Yet, even here, there is a character that exploits this intimate association.

The cuckoo catfish, as its name suggests, places its eggs in another's nest, but this nest is a little bit different, so the mother catfish watches closely the courtship and mating rituals of her potential host. The cichlid mother lays her eggs a few at a time and, as she circles around to gather them in her mouth, the male fertilises them. The behaviour is repeated many times until she has up to 100 eggs in her mouth… not all, however, may be hers.

When the cichlids begin their delicate courtship dance, their colour transforms from dull beige to flashy, bright stripes, which is irresistible to the catfish. This is when the deception begins.

During the cichlid courtship, the female cuckoo catfish darts in, eats a few of the host's eggs, and deposits a few of her own. These alien eggs are immediately fertilised by the male catfish following close behind. The eggs – both the mother's and the catfish eggs – are sucked in. Once in the mouth of the host mother, the catfish eggs have the advantage. The cichlid eggs typically hatch in up to two weeks, but the catfish fry emerge between two and four days. They feed on the cichlid eggs. The young catfish are also bigger

Below
The catfish move in, removing their hosts' eggs and replacing them with their own.

than the young cichlids. They can be 2.5 centimetres long, which might not sound much until you learn that the mother is only 10 centimetres long. The invader's fry feed on all the other cichlid babies, and when the cichlids are gone, the young catfish turn on each other. The nastiest youngster survives, which leaves a monster to emerge eventually from the mother's mouth.

The cichlid mothers do have a simple strategy to thwart the cuckoo catfish. They simply spit out all the eggs and start again. Unfortunately, they sometimes do so when there are no catfish eggs in their brood. Some kind of mistrust exists that just the presence of catfish in the area will cause the cichlids to react in this way. Whatever the reason, it is a bonus for evolutionary biologists.

Opposite top
The parent catfish eat many of the cichlids' eggs, leaving the way clear to deposit their own.

Opposite bottom
The catfish embryos will hatch before the cichlids so they can feed on the cichlid eggs, and then they turn on each other.

Above
Having eaten all its siblings, the baby cuckoo catfish is safe inside the mouth of its surrogate cichlid parent.

Arrival of the flame birds

A few of East Africa's lakes are not filled with freshwater, but are 'soda lakes'. Lake Bogoria is one. Its water is alkaline and evaporation causes it to be so salty that only specialised organisms can live in it. There are hot springs and geysers along its edge, yet despite its hyper-salinity, cyanobacteria and rotifers thrive here, and they are food for one of the world's largest populations of lesser flamingos, the smallest species.

The birds do not live here permanently, but just come to feed. They are filter feeders, dipping their upside-down beaks into the water and filtering out mainly the cyanobacterium *Spirulina*, an organism known for being a rich source of vitamins, minerals and proteins. It colours the waters blue-green, but one of its biopigments causes the flamingo's feathers to turn pink.

Breeding sites for flamingos are elsewhere. One of their main nesting areas is Lake Natron in Tanzania, where the highly caustic water deters many predators from approaching and taking chicks. There is also mounting evidence to suggest that East African birds migrate all the way to Botswana to nest. They head for the Lake Makgadikgadi Salt Pans, but only when they are flooded. How they know when to go is a mystery.

Little is also known about their lengthy migrations, including what routes they take and how far they fly in one hop. It was once thought they migrated only at night, but it is now clear that they also migrate during the day, flying extremely high to avoid predation by eagles.

With breeding over, they return to Lake Bogoria and other Kenyan lakes. At one time, most went to Lake Nakuru, but serious flooding in 2012 saw them switch destinations. Now they go to Lake Bogoria, where they arrive when the water level is low, numbers peaking between June and September. The lake is relatively predator free, there is little competition with other bird species, and the cyanobacteria blooms as the salts are concentrated at low water, so the flamingos gather in huge

numbers. A census in 2018 revealed a staggering 1.3 million birds. They turn the shallows pink! When Jo Haley and her film crew arrived, however, it was not the mass spectacle they were expecting. The birds were in small groups scattered about the lake. The only thing for it was to wait and watch what they did.

'We noticed that, as the morning progressed, the flamingos headed for a freshwater spring at one end of the lake. Bogoria is so alkaline for them that this was like an oasis in the desert. They gathered around the spring, preening and resting in a sea of pink, the spectacle we had come for.'

The flamingos followed the same routine each day, and they flew so fast that the drone from which the production team were filming barely kept up with them: they were flying at speeds in excess of 50 mph. Landing, though, was not so elegant.

'It was strange watching them fly so gracefully, but then look so clumsy on their spindly legs. Landings were a real challenge. Many had broken or injured legs. Marabou storks patrolled the colony, waiting to finish off the weak, and the entire flock dispersed in panic whenever a fish eagle flew overhead. The eagles perched on dead trees, and we found a couple of flamingo carcasses beneath their perches. It was astonishing that these raptors could take down a bird so much larger than itself.'

Above
Lake Bogoria is packed with the cyanobacterium *Arthrospira*, food for the visiting flamingos. Using their tongues, they pump water through their inverted sieve-like bills and filter out the tiny organisms.

Opposite
The flamingo's red or pink colours come from pigments in its food.

Overleaf
At times, Lake Bogoria is a temporary home to one of the world's largest aggregations of lesser flamingos.

Behind the Scenes
Successes and disappointments

The part of Africa with the most water happens to be the region with the greatest biodiversity. The Congo Basin straddles the Equator and contains huge tracts of undisturbed tropical rainforest and swamp forest. It is hot and humid all year round, with two wet seasons delivering heavy rainfall. It even controls its own weather due to transpiration by all those trees. On the global stage the forests influence rainfall patterns in the North Atlantic, and they act as a 'carbon sink', locking up carbon that would otherwise form the greenhouse gas carbon dioxide. They are also home to the critically endangered western lowland gorilla, the smallest of the gorillas and the subspecies you are most likely to see at the zoo. It was this animal that the production team had come to see in the wild, but getting to the location was an adventure in itself.

The journey took the best part of a week to reach a remote jungle camp: planes, cars, ferries, wooden canoes and finally the team and their equipment had to be taken through waist-deep swamps. When they arrived, they were met by people from the Biaka tribe, who led them through the jungle, watching out for dangerous forest elephants, honing in on gorilla tracks, and ensuring they did not get lost. It was all worth it, at least at first, as Jo Haley discovered.

'On our first morning, the trackers found the gorillas within two hours. The silverback was up in the treetops, much higher than I expected. Every time he moved, the trunk and branches swayed precariously under his enormous

Below
The western lowland gorilla is the smallest gorilla subspecies, but still a powerful primate.

Opposite
Gorillas are mainly vegetarian, eating a balanced diet of fruit and foliage. During the wet season, when it's abundant, they binge on fruit.

weight. Then, he shimmied down and stood a few metres from us. He was very intimidating, even more colossal and muscular than I had imagined. He stared at us inquisitively, but we had to avoid his gaze so we weren't seen as a threat. I vividly remember his odour: he was oozing testosterone, which lingered in the air long after he had passed.'

The team spent a couple of weeks with his troop, in which they could see that each gorilla had its own personality. Some were shy, but the youngsters were very curious. One was fascinated by his reflection in the camera lens, and Jo felt they were all incredibly gentle.

'I never saw a hint of violence amongst them. Even the silverback was placid and mild-mannered – a real gentle giant.'

The swamp forest, though, was something else altogether. It was one of the toughest shoots that Jo had been on.

'It may seem an obvious thing to say, but the jungle really is hostile – fetid, waist-deep swamps, razor-sharp saw-toothed vines, marauding hordes of army ants, venomous snakes and clouds of sweat bees – and then we were regularly legging it as fast as we could from really aggressive forest elephants, but it was all a part of everyday life deep in the Congo.'

The thick, almost impenetrable vegetation also made filming extremely difficult.

'The gorillas spent most of their time in the trees or in incredibly dense vegetation; so getting a clean shot was tricky. We would spend hours trudging through swamps, cutting down vines, and lugging half our weight in camera equipment, just for a glimpse of a gorilla.'

And, in those conditions, the sophisticated digital cameras did not stand a chance.

Above left
During the dry season, when fruit is scarce, the gorillas eat more herbs, leaves, roots and bark.

Above
Lowland gorillas climb trees, up to about 15 metres above the ground. At midday, they often rest in a day nest and at sunset they make and retire to a night nest.

'The oppressive humidity, being dragged through swamps, and sweat bees flying into the works killed the main camera on day two. Even the spare was touch and go.'

It was at this point that the team called it a day and relocated to a large clearing with open water pools in the forest known as Mbeli Bai, the base for a long-term study of western lowland gorillas that began in 1995. The bai is also a magnet for forest elephants that come here to eat clay and the minerals it contains, essential for a healthy life. The gorillas join them. The film crew could observe them all from a raised wooden platform beside the clearing, and if they were there at night, then they had to stay until morning.

'The forest elephants are most active after sunset, so we were warned not to come down under any circumstances… even to go to the toilet!'

Which was all well and good under normal circumstances, but a couple of nights in, things became particularly uncomfortable.

'We were filming elephants in the bai at sunset, when we heard very loud gunshots. Whoever was out there was very close. It was quite a shock, because we were miles from civilisation. "It must be poachers," we thought. Was it a warning shot aimed at us or had they killed an animal? Either way, we were sitting ducks on top of the platform. We were faced with the choice of staying put, knowing that armed poachers were nearby, or break the one rule we had been given and escape through the jungle. After a quick team meeting we decided to make for the nearest village, about an hour away.

'We grabbed essentials – passports, water and the footage we had shot – and began to dash through the forest. Without our trackers to see or hear elephants, it was slightly nerve-wracking, especially as the light was fading. We were soon in complete darkness, but eventually we made it to the camp

Above
The dense foliage made filming difficult for much of the day, but during siesta the gorillas were easier to reach.

unscathed, although we had a sleepless night on a wooden floor, with a friendly rat for company. In the haste to get away, we'd left behind our mozzie repellent, and so we were bitten to shreds. But, the worst of all was the fear that the animals we had been getting to know had been harmed in some way.'

The next day, an anti-poaching patrol arrived to investigate the shooting. They had travelled through the night to reach the camp, and when they searched the bai they discovered an elephant had been killed. It was the first poaching in 25 years. The production team had witnessed a serious turning point in Mbeli Bai's history.

It also meant filming was over. The presence of men with guns had spooked the gorillas. There was little chance they would show themselves for many days, even weeks. It was time to explore plan B. Director Claire Thompson and her production team headed to West Africa. Their destination was the Taï Forest, one of the last remnants of primary tropical rainforest on the Côte d'Ivoire. Living there are West African chimpanzees, another critically endangered subspecies. As omnivorous frugivores, the chimps are surrounded by plenty of food, but, in the dry season, there are a couple of favourites that are less easy to get at – coula and panda nuts. To open them, they make and use tools, and recent research has shown they have an extensive toolkit and not only to eat nuts.

The chimpanzee cracks the nut by pounding it with a large stone or a stout branch, and while most of the kernel is removed with the teeth, lips or fingers, there are always little bits left that are extracted by probing with a small stick. Short twigs are also used to 'fish' for ants, honey or marrow. The stick is poked into the entrance of an ant's nest and the soldier ant guards

Opposite
For its first two years, an infant gorilla puts on weight at twice the rate of human babies, and it is likely to depend on its mother for up to five years.

Below
In the wild western lowland gorillas live 30–40 years, while in captivity they might reach the ripe old age of 50 or more.

grab it. The chimpanzee then carefully hauls them out, sweeps them off with its lips, and swallows them. Similarly, the stick is pushed into the honeycomb of wild bees and coated with honey, and, after hunting monkeys for their meat, sticks are used to extract the marrow from long bones and pieces of brain from skulls.

Tools are also fashioned to fit the task. The ends of twigs are bitten off, for example, to shorten them, in order to extract bone marrow or the fiddly bits in nuts, and they are kept longer and thicker for ant and honey dipping. Leaves and bark are stripped away so the stick does not snag, and its end might be sharpened using the teeth. A large branch used to crack nuts is smashed against an exposed root to shorten it; alternatively, the chimpanzee stands on it and pulls it forcefully upwards to break it into two.

Filming the chimpanzees' nut-cracking behaviour was not as easy as it sounds. In their search for food, the troop would move many kilometres in a day and through the thickest jungle.

'Trekking after them, carrying heavy backpacks full of cameras, batteries, water and tripods in a rainforest with 100 per cent humidity was extremely physically demanding,' recalls Claire. And, due to our genetic closeness to chimpanzees, humans can pass on diseases, such as the common cold, which could prove fatal, so we were required to wear face masks when we were near them, making breathing even more difficult.

Below
Chimpanzee hammering tools can vary from large branches to heavy boulders.

Opposite
A baby chimpanzee watches its mother very closely in order to learn how to make and use tools and to obtain hard-to-get-at foods, such as the protein-rich kernels of nuts.

As we stumbled at speed behind the group, vines would snag us, pulling us backwards, forwards and tripping us up or slicing through our clothes and skin. The chimps would saunter elegantly ahead of us, navigating their home with such ease and confidence.'

So, keeping up with them, let alone filming was exceedingly difficult, but there was often a respite in the middle of the day. Watching the family take a siesta was remarkably familiar.

'All the hardship was worth it, just to get to know their individual characters: the boisterous, the clowns, the shy and the curious. It was like watching a soap opera unfold in front of us. A constant cacophony of screeching males jostling for the top spot contrasted with intimate grooming sessions filled with such gentle care.'

The midday rest period was also an opportunity to touch base with the office back in the UK, but that was a long-winded process. Claire had to type in a message and hoist her phone to the treetops in a bucket, wait for it to receive the message then repeat the process all over again. However, when they could get down to filming, Claire and her crew focused on 43-years-old Perla and her five-year-old daughter Pegatta.

'Perla is one of the finest nut crackers in the whole group, and a stand-

Above
Pegatta watches Perla cracking nuts. Shortly afterwards she went away and practised the technique herself, albeit with a much smaller stone!

out moment for me was watching Pegatta diligently practice her nut-cracking skills under the watchful eye of her mother. It'll be another five years before she'll fully master the technique, but she showed such enthusiasm and tenacity in her pursuit, that it was impossible not to admire her dedication.'

The tools used by the chimpanzees are sometimes made in advance, rather than during the task, which indicates that they show some understanding of the relationship between objects, enabling them to manufacture specific tools for a particular job. To crack nuts, Claire noticed Perla not only had her own favourite hammer stone, but that it might also be shared around.

'Stone hammers are heavy and are highly prized commodities. We would see queues forming where a chimpanzee would wait patiently for the other to finish and the hammer to become available.

'Interestingly, anecdotal evidence suggests that this behaviour was first discovered when the colonial French military heard banging noises coming from the forest. They thought it was a rebel army forging new weapons.'

And, to think, in the not-so-distant past, that humans were conceited enough to think we were the only species to make and use tools. Science has certainly come a long way!

Above
Chimpanzees prefer fruit above anything else, but will eat nuts, seeds, leaves, flowers, bark and resin, as well as honey, bees, ants, termites, birds and their eggs and monkeys.

Europe

Producer: Giles Badger

Introduction

When seen from space at night, almost the entire continent of Europe, including European Russia, is flooded with artificial light. There are a few large dark spaces, most of them in the north and east. These are the wilderness areas. Some are protected as national parks or nature reserves, while others are just patches of green amongst hectares of grey concrete, so Europeans can still keep in touch with nature, no matter how fettered it is. They can listen to wolves howling at night, encounter a bear on a forest path, watch seabirds plunging for fish, and they can even spot a wild hamster in a city graveyard.

Europe, however, is a crowded place, the third most-populated continent, after Asia and Africa, but with people confined to a significantly smaller area – and there are more arriving all the time. Immigration is not new, though. Modern humans arrived in Europe from Africa about 80,000 years ago, but it was not until after the last Ice Age, about 10,000 years ago, that people began to have a major impact and shape the continent. Its rivers were one reason. As the ice sheet retreated, ancient peoples migrated, settled and farmed along Europe's many rivers, which meant the continent filled up with people and it was changed radically, leaving less space for wildlife. Agriculture, silviculture and urbanisation throughout history have altered the face of wild Europe, but wildlife has found ways to take advantage of what living alongside humans has to offer, albeit hidden away, lest it be shot, hunted, or run over! And, with re-wilding now taking root in many parts of Europe, in which core wilderness areas are restored, wild space created and key species reintroduced, European wildlife might well have a chance to experience Europe as it once was, at least, in some parts of the continent.

Right
An aerial view of the River Koiva on the border between Estonia and Latvia, in Northern Europe.

Baby kidnappers

The Barbary macaque gets its common English name from the land of the Berbers – the Barbary Coast of North Africa, where the largest populations of this Old World monkey still exist. How they came to be on the Upper Rock of Gibraltar, on the European side of the Mediterranean, is anybody's guess. The macaques could be a relict population from one that was more widespread across Southern Europe and North Africa in the distant past, or people could have carried them there. If DNA analysis is anything to go by, monkeys from North Africa must have been imported from time to time to top up the numbers, perhaps to quell the superstition that if the macaques leave the rock the British will go too.

Whatever its place of origin, the Barbary macaque is unique for several reasons: it is the only species of macaque living outside of Asia, the only primate in the northern area of the Sahara, and, aside from humans, it is the only primate living semi-naturally in Europe, where it is also known as the 'Barbary ape', because it has no tail.

Finding food on the crowded peninsula is not a problem because Gibraltar's macaques are fed daily with water, fruits and vegetables, care of the Gibraltar Ornithological and Natural History Society. The Gibraltar Veterinary Clinic also monitors their health, so they are extremely cosseted monkeys. There are currently eight troops, and, with such regular meals, those troops tend to be big, up to 50 monkeys in one case. In macaque society, however, much like in human society, there are the haves and the have-nots – the high- and low-ranking individuals. Director Kiri Cashell and her film crew chose to follow a 'have-not', probably the lowest-ranking female in her troop.

Opposite
The mother and baby Barbary macaque may be at the bottom of the troop's hierarchy, but they can rely on each other for support.

Below
Gibraltar's macaques live on the Rock. While Barbary macaque populations are declining in North Africa, Gibraltar's are increasing.

'The troop we filmed has its territory in one of the most popular tourist spots, where the monkeys use roads and walls as their own highways. Their high-rise home was a spectacular place. The sounds of the city filtered up from below, and, on a clear day, you could see Morocco in the distance. It was an unusual place to be filming wildlife.'

It was July when Kiri first visited the Rock, and the low-ranking female had just given birth. She was at the bottom of the pecking order. Both the mother and her baby were outcasts, forced to live on the fringes of the troop, so she didn't have much in the way of food, but at least she and her baby had each other. Though she seemed a good mother, her offspring sometimes had what can only be described as a 'hellish' infancy, and it started when it was just one week old.

Low-ranking males often kidnap babies, particularly those of low-ranking females, to help them climb the social ladder. When two males fight over a potential mate, the loser often presents the victor with a kidnapped baby macaque as an appeasement, but this time the abductor was a young female.

'She seemed to want to play at being a mother and, while we were filming, she saw an opportunity. The youngster was playing on the ground and exploring away from its mother, when she grabbed it. The baby started to cry and the mother tried to get it back, but we think the captor was of a higher rank. Everything in macaque society is determined by rank, so the low-ranking mother could only look on and watch what was happening to her baby.'

What followed was heart-stopping. The pretend mother, gripping the hijacked baby, ran amongst heavy traffic and, being inexperienced, she did not hold it properly. She was dragging it around, and then scaled a tall cable car pylon, but the real mother was close behind. On the top of the cable car, she retrieved her baby, and she did it in a clever way.

'She groomed another monkey in front of the higher-ranking female, and the urge to join in was so strong – because grooming is one of the most important social interactions for macaques – that the baby was able to crawl back to its real mother. The young mother may have been of low rank, but her intelligence gave her a distinct advantage in Gibraltar's Barbary macaque society.'

Below
The low-ranking mother manages to retrieve her baby by instigating a mutual grooming session to distract the abductor.

Grave robbers

Europe's second-largest cemetery is the Wiener Zentralfriedhof in Vienna. With 330,000 gravesites, it is the final resting place of Ludwig van Beethoven – Grave No. 29 in Group 32a. He was buried in and exhumed from two other sites in Vienna before this became his final resting place. Vienna is one of Europe's greenest cities, though, and the Wiener Zentralfriedhof is also home to the living. The European hamster is a resident, a larger version of the Syrian hamster commonly kept as pets. At any one time, up to 1,035 individuals live on the 238-hectare site, which also includes a spectacular grove of fifty English oaks.

The Viennese animals have opted to live in the cemetery because people are frequently about, and so fewer predators disturb them. It was something Kiri Cashell noticed.

'The graveyards were tranquil havens in the heart of the city, but they were also relatively busy with gardeners, walkers and people visiting graves. In the autumn, the hamsters are busy collecting food for their winter larders, so you could see them scuttling about everywhere.'

In fact, they have even become partly diurnal to avoid the nocturnal foxes. It means they're much bolder than elsewhere, using the gaps between graves like motorways, and the visiting mourners unknowingly bring with them ample supplies of food.

Flowers left at gravesides are a delicacy and hamsters will fight vigorously with others of their own kind for the right to pillage them, but it is a fondness for candle wax that can get them into real trouble. People leave candles in glass jars. When the flame is extinguished by the wind, the hamsters make a beeline for the jar. Pushing its head right inside, a hamster will nibble away at the wax and store it in its elastic cheek pouches, in order to carry the hoard back to a special underground food storage chamber… but there's a problem. The bulging pouches can hold up to 80 grams of food and, as they expand, the hamster's head becomes stuck in the jar.

'There was always a lot of wriggling around and pulling at the candle with their paws,' observed Kiri, 'but eventually, they do manage to get their heads free.'

Baby cave dragons

The olm or proteus is an unusual species of aquatic salamander that lives permanently in the dark. One of its homes is the labyrinth of caves underneath Slovenia – the so-called 'basement of Slovenia'.

Local people refer to the olm as the 'human fish' on account of its pink skin colour, and its strangeness has given rise to all manner of legends.

Naturalist Johann Weikhard von Valvasor first described the olm in 1689. In his *Glory of the Grand Duchy of Carniola*, he tells how local folk believe the creatures to be baby cave dragons. They are certainly, let it be said, unusual.

Olms are eel-like in shape, up to 30 centimetres long, with a flattened tail edged with a flimsy fin. Living in darkness, their eyes are undeveloped and covered by a layer of skin so they are functionally blind, although they can sense light and will shun it. By contrast, their senses of smell, touch and

hearing are acute. They also have sensors in the snout capable of detecting weak electric currents, and so can probably not only detect the electrical activity in the muscles of their prey, but also sense the Earth's geomagnetic field and maybe use it to work out where they are in the darkness of the caves. They retain larval features, such as feathery external gills, and their thin layer of pink-coloured skin is translucent, so their abdominal organs show through. The pink colour comes from the red blood cells coursing through their arteries.

The temperature of the well-oxygenated water in their caves is relatively cold at 8–11°C, and river levels rise and fall throughout the year, but food is not always readily available. Olms eat freshwater shrimps, aquatic snails and the occasional insect, which they swallow whole, and they binge eat when food is plentiful, storing fats and sugars in the liver for when times are hard. They can also slow down their metabolism and curtail normal activities;

Above
The speed with which olm embryos develop to hatching depends on the water temperature: at 10°C it takes 140 days, while at 15°C it is only 86 days. It will then have another 14 years before reaching maturity.

and, if push comes to shove, they can reabsorb some of their tissues. In fact, generally, these amphibians move very little, usually to feed or to mate, the latter once in every 12.5 years on average. It's as if time stands still.

There are no predators down there, so their life is effectively stress-free. As a consequence, experiments have shown that they can survive for up to 10 years without food, and one study has estimated that they have a maximum lifespan of 100 years, although 68.5 years is thought to be the average, the longest of any amphibian. IUCN lists the olm as 'vulnerable', but salvation is at hand. Olms are being bred in specially built aquaria located in the cave system at Postojna. Here they are fed regularly and protected, resulting in a successful captive breeding programme. The future for baby dragons is beginning to look brighter.

Above
Locals call the olm the 'human fish' on account of its flesh-like skin colour. It has eyes but they are underdeveloped, so the animal is functionally blind.

Tisza flowers

In mid to late June each year, the most extraordinary thing happens on the Tisza River in Hungary. It is the time that giant mayflies appear. For the previous three years, they had been out of general view as white aquatic nymphs that moult many times on the bed of the river, but come the spring at the start of their fourth year, they emerge en masse. In the space of a few hours, millions of winged mayflies, the largest known species in Europe – 8–12 centimetres long – dance and mate, but do not feed. They fly, males moulting one extra time after acquiring functional wings, before fighting for the right

to mate, and then, swooping close to the water, engaging with the opposite sex. Several males crowd around a single female, the mating frenzy known as 'Tisza flowers'. The insects have just three hours to complete their life cycle and maintain the future of the species, and it kept producer Giles Badger and his production team on their toes too. They travelled across the entire country chasing swarms.

'The river gradually warms as it flows, so swarms start along its lower courses, and then gradually occur upstream. Each day, a different river bend explodes with life, so we had to second-guess where the swarms would appear.'

The film crew had little warning. For hours only the river flowed slowly past their boat, but then a lone mayfly broke free from the surface.

'Within fifteen minutes we were engulfed by a swarm that was so thick, we could barely see through it. Mayflies landed on our clothes and moulted and then we were surrounded by the clatter of mayfly wings.'

The female becomes sexually mature when she moults her last skin and, after mating, the males who clustered around her die and she begins a journey upstream. Joining hundreds of thousands of other females in a single swarm,

she flies at a height of 5–10 metres before she drops down onto the water and releases her eggs into the river. The location is precise, for the eggs drift downstream and gradually sink, ending up in the bottom of the deep pools from which the parents had just emerged. After about 45 days, the nymphs hatch out, just as their mother did several years before. They hide on the riverbed for three years before they become adults themselves and join the spectacular mass emergence; if pollution doesn't affect them, that is.

'We were told that the swarms were once much bigger than they are today,' reflects Giles, 'and we saw a lot of human debris washed down by the river. The river has not only been a thoroughfare for people, but also a dumping ground for their waste and this has affected mayfly numbers.'

Danube pirates

Tisza's waters wind their way across Europe and eventually reach the sea in the Danube Delta, another of Europe's great natural wonders.

'I was amazed that a wilderness of this scale existed in Europe,' says director Jo Haley, 'an unexpected slice of paradise for nature lovers. We were there at the summer solstice. Each morning, just before dawn, thick mist would hang over the waterways, followed by beautiful reflections of the sunrise in the lakes. Then, the place came alive – in the golden light you'd realise that it was absolutely teeming with birds.'

And, it was the birds – one species in particular, the great white pelican – that Jo and her film crew had come to see. During the spring and summer, this vast maze of marshes, lakes, ponds, channels and streams is home to millions of breeding birds, including more than half of the world's great white pelicans.

Pelicans tend to be territorial at breeding time but, when away from the nest, the great white pelican is highly sociable, especially at feeding time. Around midday, the birds gather into small flocks ready to fish together. They line up in a horseshoe-shaped formation and drive shoals of fish into the shallows. Then, at some unknown signal, all the pelicans thrust their heads

and enormous beaks underwater simultaneously, scooping up the fish; but there is, it seems, an easier way to make a catch.

The pelicans take to the air and use thermals to soar upwards and scan the delta for signs of another species of water bird – the pygmy cormorant. It only takes a couple of pelicans to land close to a group of cormorants and birds coming flocking in from all around. The expectation was that this mixed fishing party would see the two species cooperate. As the pelicans plunged their heads below, the cormorants dive down and snatch what they can, and probably inadvertently drive fish towards the pelicans' huge bills. Working together like this, you'd think each would catch more fish than if they were fishing alone, but studies indicate otherwise. It was this cooperative feeding behaviour that Jo had come to capture on film but, while they were filming, she noticed that there was something else going on, something a little more sinister.

'The action seemed to occur when the morning's fishing had calmed down, and you could almost guess where. The birds lined up, with the cormorants fishing at the front, and the pelicans following along behind; but when a cormorant dived and caught a fish, the mood changed. As soon as it surfaced, two or three pelicans would lunge at it. Instead of working alongside the smaller birds, as expected, they would bully them and then rob them.

Below
Great white pelicans might look like gentle giants, but they are not as innocent as they might seem.

'The cormorant would try frantically to swallow its fish, but any hesitation and the pelicans would grab it by the neck. Between being choked and the shock of being grabbed, the cormorants were forced to release their catch, straight into the mouth of the nearest pelican.'

Jo noticed that the pelicans at the very front of the formation seemed to be the worst culprits, and they became more aggressive as the fish numbers dwindled.

'I think the pelicans had found an energy-saving way to feed: they let the agile cormorants chase after the few remaining fish and then thieve an easy meal. Why the cormorants put up with it, I've no idea. They were doing all the hard work, and didn't really need the pelicans at all.'

The pirating behaviour Jo and her crew filmed was totally unexpected. There are few references to it in the scientific literature. They went to film a sequence about one species cooperating with another, but it turned out to be about robbery with violence!

Opposite
Pelicans have the largest bill of any bird. The Dalmatian pelican is the other pelican species in the Danube Delta.

Below
Pelicans invade a group of cormorants, making them regurgitate their hard-won catch for the bullies to purloin themselves.

Overleaf
When they are not stealing, great white pelicans gather in huge feeding flocks to corral and grab fish in a synchronised plunge.

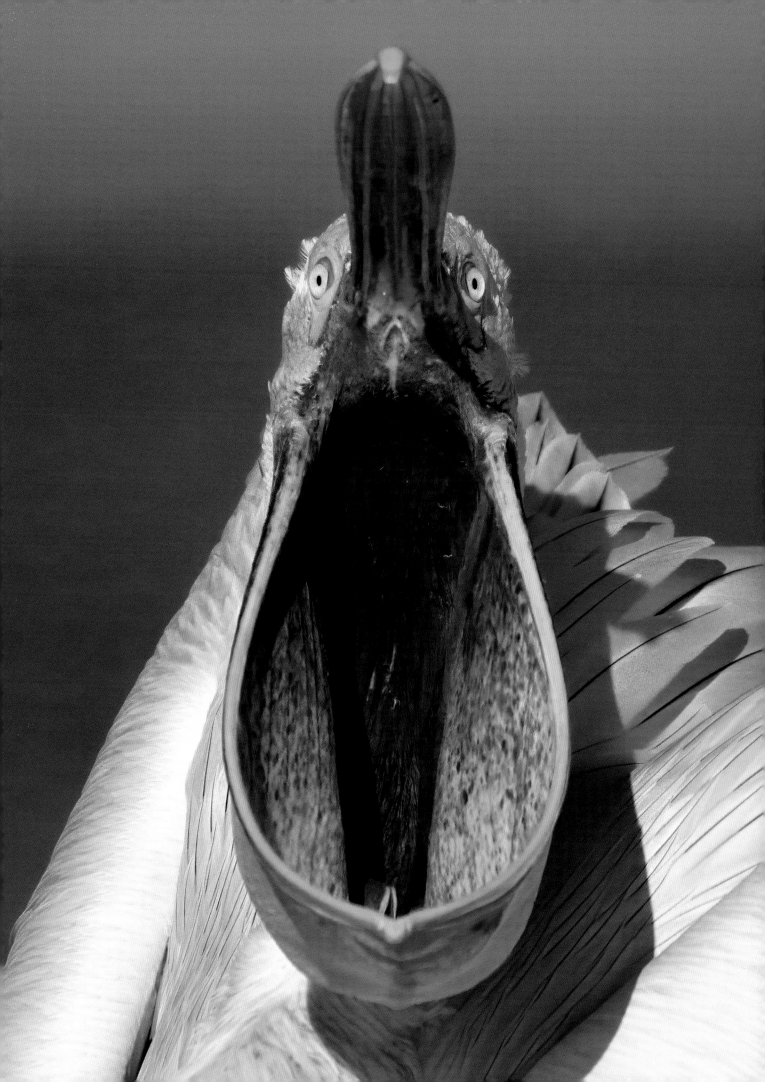

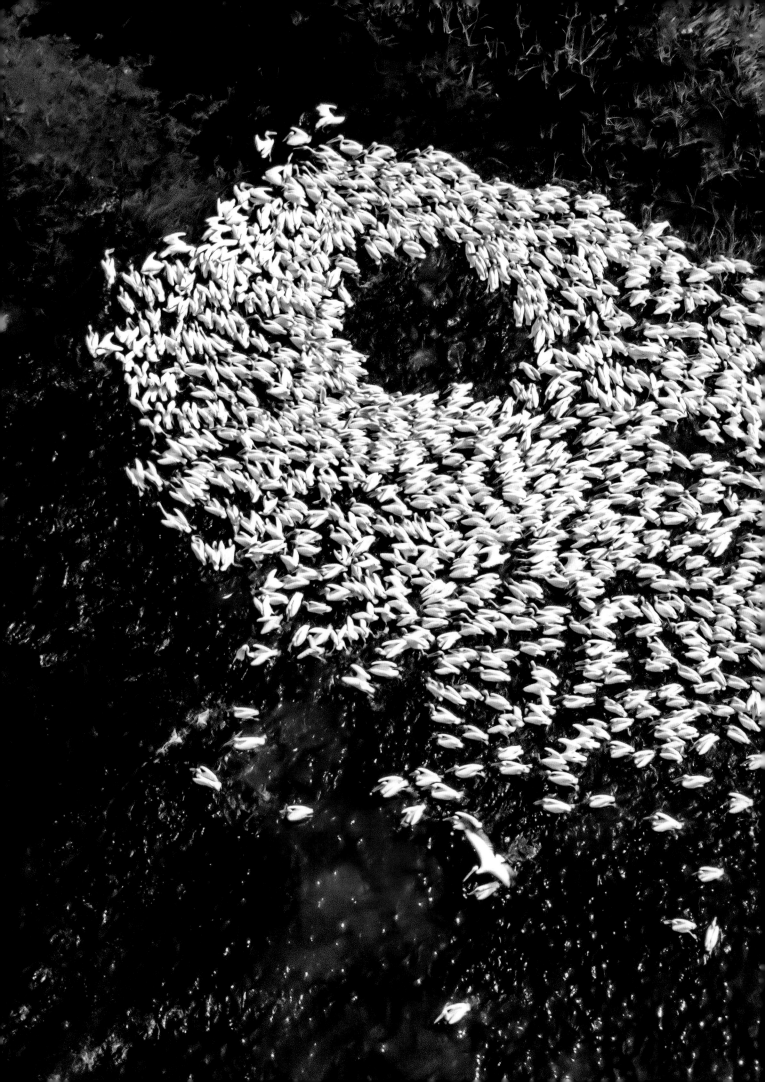

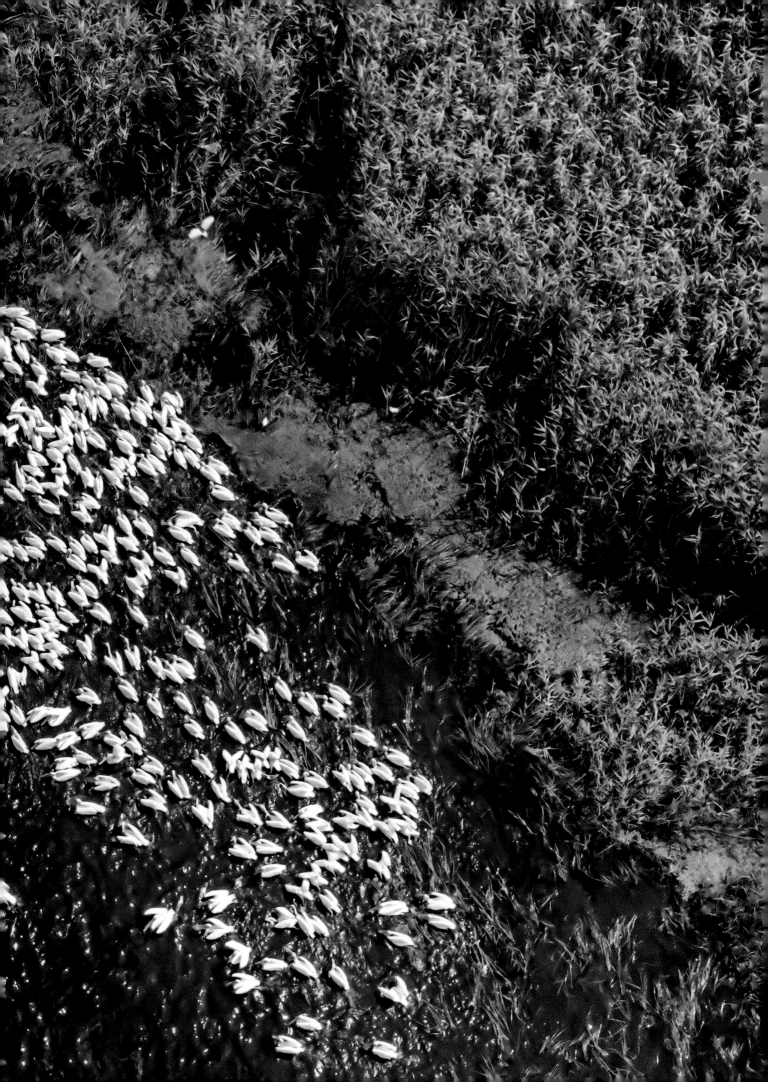

Head-bangers

True relics of the last Ice Age, the musk oxen of northern lands were around at the same time as woolly mammoths and cave bears. They have taken ice ages, disease, human hunting and a close call with extinction in their stride, having been on this Earth three times longer than we have. About 10,000 years ago, they were to be found all across the Arctic, but they were hunted mercilessly. They had been extirpated in Europe and Asia by the mid-1800s, and animals living there today were reintroduced from Greenland and Arctic Canada. After the Second World War, for example, musk oxen were released in the Dovrefjell-Sunndalsfjella National Park in Norway, and it was the descendants of these animals that Jo Haley and Jack Delf, together with cameraman James Ewen, had come to film.

'It was rugged, bleak and desolate. Most days we were fighting against howling winds, pouring rain and biting cold under bruised skies,' says Jo. 'It fitted the sequence perfectly: primordial beasts in a harsh and brutal land.'

The primordial musk ox gets its name from the odour the male emits to entice females during the breeding season. However, despite being labelled 'ox', the animal is more closely related to sheep and goats than cattle, a so-called 'goat-antelope'. Its closest living relatives are the goat-like gorals of Asia. When seen on television, the large head, shoulder hump and thick, double-layered coat make the musk oxen look massive, at least the size of cattle, but actually they are quite small, half the weight of a bison and about 1–1.5 metres tall at the shoulder. Both sexes have strong, curved horns that meet at the centre of the forehead to form a large and imposing 'boss', male horns being larger and heavier than those of females.

The long, outer guard hairs of their thick coats reach almost to the ground. It gave rise to the Inupiaq-speaking Eskimos calling the animal *itomingmak*, meaning 'the animal with skin like a beard'. Wool from the dense inner coat, which is lost in summer, is known as *qiviut*. People collect it, as it is highly prized for its softness and insulating properties, and it is said to be the most rare natural fibre in the world.

Musk oxen live in herds in which males and females have their own hierarchies, with older, mature animals lording it over the youngsters. Older males will rush at younger animals and butt them in the flanks. They'll also roar, swing their heads and paw the ground, and occasionally these dominant bulls treat less-dominant males like females. They kick the foreleg, as they do when courting a female. They'll also engage in mock copulations and sniff at the smaller male's genitals, all designed to remind them of their lowly position in the herd.

The herd does not occupy a territory, but the animals scent-mark their trails, and mature oxen tend to grab the best pastures and foraging sites. They eat lichens, mosses and grasses and, in winter, mainly Arctic willow. They are not particularly good at digging in the snow, so herds tend to seek out higher ground in winter, where the snow is thinner.

Opposite
A musk ox shakes off snow after a snowstorm in Dovrefjell-Sunndalsfjella National Park, Norway.

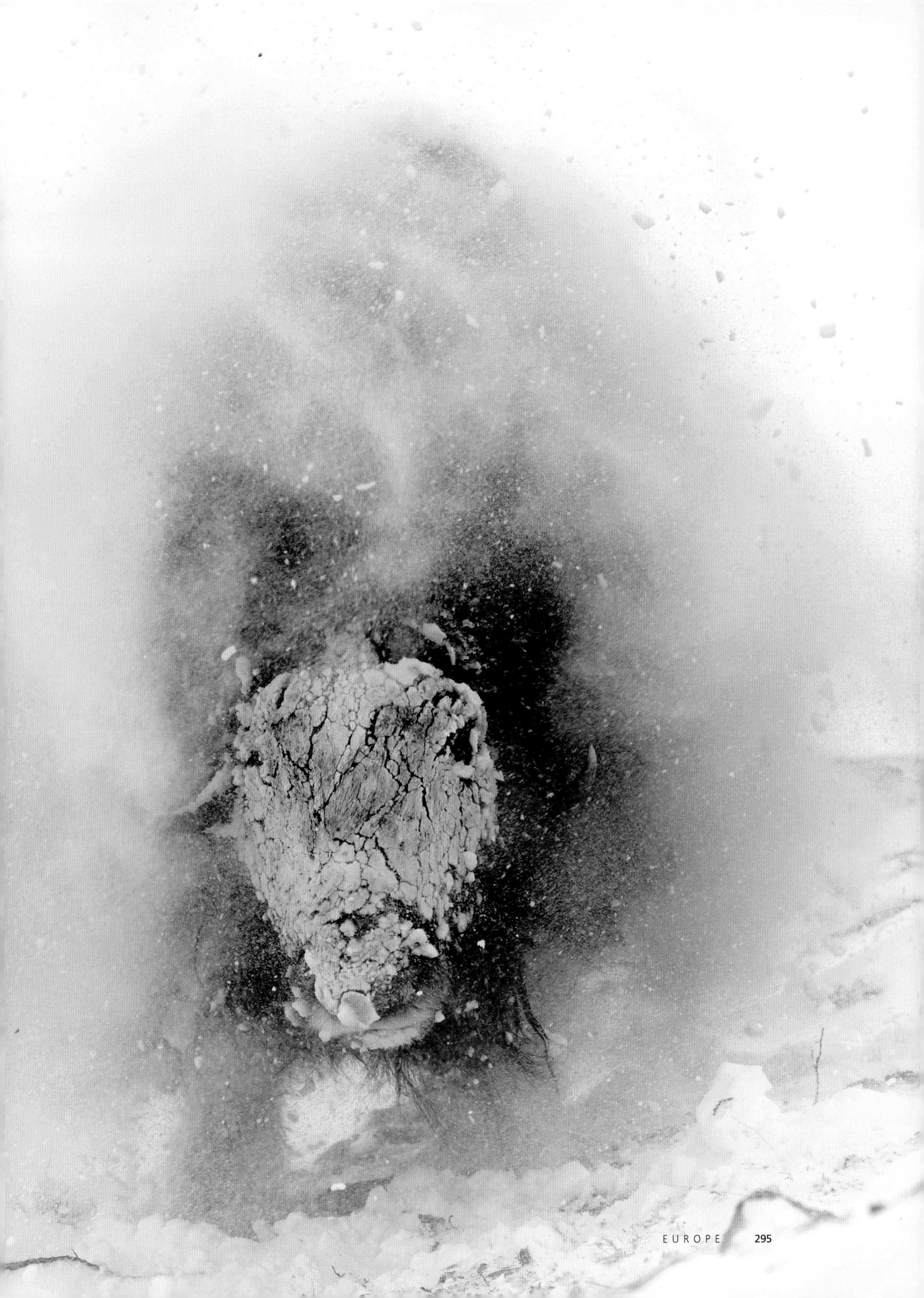

Below
Two male musk ox
come face to face in
Dovrefjell-Sunndalsfjella
National Park.

The rut is in late summer and early autumn, and just before things begin to liven up, males have been seen to jump around like tormented animals, rushing about in circles, jumping on the spot, and splashing through ponds, all the while sizing up a large rock and then gently head-butting it while practising for the rut. They'll also spar half-heartedly with other pretenders before the real thing.

During the rut proper, males that appear equally matched square up to each other. Fights occur very quickly, as Jo discovered.

'We noticed a lone bull walking quickly and with purpose towards a harem. The harem male detected the intruder and, turning towards him, started to run. He showed no signs of slowing down, picking up speed until he crashed into the horns of his opponent.'

On another occasion, Jack and James were on the moor, each about a kilometre apart, scanning for activity, when their guide appeared suddenly on the horizon shouting: 'Fight, fight, fight!'

'We raced over,' Jack recalls, 'and there were two huge males circling one another, digging their heads into the ground and covering their horns with mud and grass.'

The two combatants then rubbed glands on their faces against their legs, and roared. Each swung his head from side to side and touched the opponent's head, both trying to gain the upper ground. Then, they backed up about 20 metres, lowered their heads and charged into each other at full tilt, smashing and then rubbing heads together before another assault.

'The noise was unforgettable,' Jo remembers. 'The deafening clash reverberated throughout the valley. Looking at the force when they hit, you were amazed how their horns could still be intact, the impact trembling through their long fur.'

In fact, the skull is heavily armoured, with 10 centimetres of horn and seven centimetres of bone protecting the brain from damage. They might keep this up for as many as 20 clashes, the winner broad-siding his weaker opponent, slamming into his side, lifting him bodily off the ground and pushing

him downhill, using the higher ground to his advantage. Eventually, the loser backs down and makes his escape. The victor might have a harem of six or seven females, which he defends against all comers. He'll chase competitors away from his herd, and he has a fair turn of speed.

The loser and the subordinate males form bachelor herds, but they are allowed back with the others if danger threatens. At the end of the summer, they all get back together again, forming large mixed-sex herds during the winter; but, while the males influence events during the rut, the pregnant females take charge for the rest of the year. They become aggressive and decide where and when the herd will travel and rest.

At Dovrefjell, musk oxen have few natural predators, but in other parts of the world, their main threat, apart from human hunters, is the wolf. Bears take calves, but a wolf pack will kill any animal that is weak, helpless or isolated from the herd. The herd's defence is to form a phalanx, each adult facing outwards and presenting their horns to the attacker, the calves hidden behind their legs. They really are remarkable survivors.

If you go down to the woods today...

At one time, there were brown bears all over Europe: wherever there were forests, there were bears. Nowadays, they are restricted to the more remote wilderness areas, where there are fewer people. Such places exist in Finland, where more than a thousand bears live in the forests, especially in Eastern Finland, on the border with Russia, where you get a glimpse of the Europe of old when trees dominated the landscape and large animals roamed.

Mating takes place in June, when it is light for almost 24 hours. The males scent-mark their territories and, should they meet a rival, their size is an

indication of their strength. Small males are wary and avoid bigger males, but if two bears of about the same size confront one another, there is sure to be a fight. The victor will follow a female relentlessly until she seems to give in due to sheer exhaustion.

The cubs are born the following winter in the mother's hibernation den. They all emerge in springtime, usually up to three cubs, rarely four, and, aside from being taught by their mother how to be and live like a bear, they must learn how to climb. Director Charlotte Bostock and her team visited the same area over a three-month period, watching cubs become increasingly confident as they got older.

Below
One stronghold for Eurasian brown bears is on the Finnish-Russian border.

'On one occasion, the smallest cub climbed too high and seemed to get vertigo. He just froze at the top of a tree that must have been about 30 metres high. He was quivering with fright, looking for a way down, but unable to move. His mother was at the base of the tree, encouraging him to come down. She must have been there for a good hour, and she eventually coaxed him down safely.'

Climbing trees like this is important for bear cubs. If a large male bear chances upon the family, the cubs can be in mortal danger. The mother will challenge the intruder, snapping her jaws aggressively, but, just to be safe, the young bears must be able to shin up the nearest pine tree.

Above, Opposite and Overleaf
Brown bear cubs play-fight to prepare themselves for adult life, but while young they are vulnerable not only to attacks by predators, such as wolves, but also male bears. Males have been known to kill youngsters. As adult males tend not to climb, the safest place for young bears is up the nearest tree.

The comeback cat

The Iberian lynx is at the centre of an ongoing conservation story that has been on a knife-edge for several years. Much of its living space has been lost in the past century, and in 2002 there were just two breeding subpopulations in southern Spain, with 94 individuals. The species was literally on the brink of extinction, the most endangered feline in the world, but there are signs that it is bouncing back.

The animal itself is a small cat, twice as long and three times as heavy as a domestic cat, with ear tufts, very short tail, and distinctive facial whiskers. It lives in the Mediterranean forests of the Iberian Peninsula. Where and how far it roams is determined by how much food is available. Home ranges can be up to 100 square kilometres, ownership advertised by scent marking with urine, droppings and scratch marks on trees, and, once an area is established, it tends to be stable for several years. Within it, a lynx can walk 50 kilometres a day in search of prey.

Food is mainly rabbits. A single lynx requires one rabbit a day to survive, and a mother lynx with a family will need to catch at least three. If rabbits are unavailable, ducks, partridges, quail and young deer and mouflon are caught instead, but, generally, rabbits account for up to 90 per cent of its diet. This has been part of the reason for the lynx's declining population. Myxamatosis

Above and Opposite
Separated by the Pyrenees, the Iberian lynx evolved separately from the other lynx populations in Europe. It has tufted ears, like all lynxes, but its facial whiskers or ruff are significantly longer. It is about half the size of the Eurasian lynx. It has a shorter coat to cope with the mild Mediterranean climate, and it has more spots than other lynx.

and rabbit haemorrhagic disease has knocked out its main food supply, and they are not the only problem. The construction of motorways and railways has split up the lynx's habitat, and illegal hunting has taken its toll. There is hope, though: Lynx are being reintroduced to locations in Spain and Portugal with some success. In the Sierra de Andújar, Charlotte Bostock had the luck to meet one close up.

'Because they are so rare and roam over large areas, filming the lynx was like finding a needle in a haystack. We opted for camera traps to help locate them, which is when I first came across an incredible male lynx named Farron.'

Farron was an unusual lynx. Instead of Charlotte searching for Farron, Farron found Charlotte!

'Unannounced, he would suddenly appear, and then sit and watch us as we worked. One day, I was checking the camera traps on my own when Farron was suddenly beside me. We sat together under the shade of a tree for about 30 minutes. It was a real privilege to be in the presence of such a majestic and rare cat, whose near extinction was our responsibility.'

One of the reasons Charlotte was filming Iberian lynx was to celebrate its return, and highlight the European Union's LIFE programme, which helped fund its recovery.

Above
The Iberian lynx is one of
the rarest cats on Earth,
so every kitten counts.

'We want to show that, in order to save a species and then protect it in today's rapidly changing landscapes, we must make space for them alongside us and give them a helping hand. In the case of the lynx, it has meant managing the land so that the lynx are undisturbed, ensuring there are enough rabbits to eat, and running a complex reintroduction programme to keep the gene pool varied. Without these measures, this species probably would not have been able to recover from the population decline.'

Latest figures show the overall Iberian lynx population had grown to 686 by May 2019; in fact, the species is recovering so well that the IUCN Red List has demoted its status from 'critically endangered' to 'endangered', with the

epitaph 'excellent proof that conservation really works'. The key thing now is to create corridors between the isolated populations so lynx from different parts of the peninsula can intermingle and not inbreed.

However, not everybody is so optimistic about the fate of the species. They point out that rabbits are still succumbing to diseases, poachers are killing animals, and there are collisions with cars and trains. Extinction, despite all the conservation effort, is still possible within a few decades; but for the moment, it does show that even on a crowded continent like Europe, there are ways in which people and wildlife can live alongside each other.

Behind the Scenes
Finding Italy's wolves

Italy's Abruzzo region, to the east of Rome, is what some observers consider to be the greenest region in the whole of Europe. Aside from the vineyards that offer us Montepulciano and Trebbiano d'Abruzzo, over half of its area is given over to national and regional parks, nature reserves, and Europe's southernmost glacier – the Calderone. It ensures that many of Europe's wild animal species are protected here. It also means that isolated rural communities have wolves for neighbours.

The Apennine wolf is a subspecies of the grey wolf, and native to Italy. At one time it was shot indiscriminately, so only a handful of wolves survived before it was declared a protected species in 1971. Conservation work, however, has been rewarded with some success. At last count, about 1,600 of them are thought to be living in Italy, with a few moving into France and Switzerland in the north. The subspecies is recognised by having a pelt that is of a grey-tawny colour, with a distinct tinge of red in summer. Kiri Cashell went in search of them, but, at first, her wild wolf hunt turned into more of a wild goose chase.

'Our first location,' she recalls,' was a wild setting, high in the mountains, but we had little luck. For over a week, we had brief glimpses of wolves, but that was all. We had to find a better place.'

Luckily, a local farmer spotted wolf kill on the road and gave the team the tip-off.

'The irony is that it was right next to the village in which we were staying. It was a charming place, one of the red-roofed, hilltop villages, not at all the place you'd expect to see wolves, so we staked out the area with thermal cameras and immediately came up trumps. We witnessed our first wolf hunt, so we relocated here.'

Below left
Abruzzo, to the east of Rome, is sometimes described as 'the greenest region in Europe', for over half of it is protected as national parks and nature reserves.

Below
In the west, Abruzzo's mountains are one of the few remaining homes of the rare Apennine wolf.

The wolves generally avoid people and hunt mainly at night when most people are tucked up in bed. Their principal targets are deer, wild boar and chamois, but they'll also take hares and rabbits, as well as eat berries when in season. Scarcity of large prey, though, means that packs are relatively small, but where deer have been reintroduced and are thriving, there can be as many as six or seven pack members.

The production team followed a small pack, and their speciality was to catch red deer. Thermal cameras, in cars parked in the car park at the edge of the village, could pick out the wolves in complete darkness. Kiri found that the pack hunted every night, and that not everyone in the village was asleep. An unexpected bunch of wolf watchers joined the team.

'We soon realised that all of the dogs in the village were our early warning system that the pack was in the area. They would all erupt at once, their barks and howls echoing up and down the valley. Moments later we would see the wolves appear in the distance. The dogs always seemed to know.'

More secretive than their American cousins, the wolves use stealth and the cover of darkness to approach the deer. They climb rapidly so they are directly above their quarry, and then each wolf must sneak up as close as it can to the nervous animals. The deer are on high alert so, to have any chance of catching them, the wolves must set up an ambush; but, with the whole herd listening out for signs of danger, the odds are stacked against the pack. Once the deer have detected them, the pack has no option but to stick with them until they calm down. It seemed to the production team that the wolves were actively chasing the deer downhill, and often the roads, which were dug into the mountainside, would cause deer to trip, fall or slip on the ice.

While the crew was with them, the pack was chased away from their kill by local sheepdogs, but the wolf's reputation as a stock killer has seen farmers hunting them and exhibiting their mutilated bodies as a macabre protest. In

Below left
The Apennine wolf is the unofficial national animal symbol of Italy.

Below
At sunset in the mountains, the film crew sets up a specialist camera that will enable them to film the wolves hunting deer at night.

Left
The Apennine
wolf in its natural
surroundings: an
ancient beech forest
on a mountain slope
just an hour and a half
from the 'Eternal City'.

Tuscany alone, wolves and feral dogs behaving like wolves have been causing an estimated 1.2 million Euros' worth of damage to livestock a year, and so farmers fear for their livelihoods. Italian wolves may be having a dramatic comeback, but it is not something with which the entire human population is comfortable, despite the wolf being the unofficial national animal of Italy. It all harks back to the legend of Romulus and Remus, the co-founders of Rome, having been reared and protected by a she-wolf... but will Italians dig into their roots and protect their wolf? Kiri discovered that there was room for some optimism.

'Our village was a sheep-farming village, with just a few hundred people living there, and many of the farmers had had run-ins with wolves. They had attacked livestock, but there was one farmer with whom we talked whose attitude was surprisingly positive. Over the years, he had lost only a few sheep because, he said, you can farm them in a way to prevent this happening. He used his dogs to protect his flock and no harm came to Italy's unique subspecies of wolf.'

Afterword
One Planet Under Siege

Since 1964, the International Union for the Conservation of Nature (IUCN) has monitored the fate of plants, fungi and animals on the seven continents and in the surrounding oceans, and it has compiled a Red List of species threatened with extinction. As of 2019, it lists more than 27,000 species that are classified as 'vulnerable', 'endangered' or 'critically endangered', which represents 27 per cent of all the species that have been assessed. About 40 per cent of amphibians, 34 per cent of conifers, 25 per cent of mammals, and 14 per cent of birds are in the firing line. It's a clear sign that biodiversity is declining the world over.

Asia has the doubtful distinction of being the continent with the most endangered species – about 3,330 – but then it also has high biodiversity, so it has a lot to lose. Indonesia alone has 12 per cent of the world's known mammal species, but it also has the highest number of endangered mammals – the Sumatran rhino being one – yet the country occupies only 1 per cent of the world's landmass. Being of such a size and divided up into many small islands, the wildlife is especially vulnerable to even minute changes in the environment; in fact, globally 90 per cent of birds that have become extinct since 1500 lived on islands, one of the latest being Hawaii's poo-uli, last seen on Maui in 2004.

A recent announcement from Birdlife International, however, revealed that of eight bird species set to have their extinctions in the wild confirmed, five are from South America, including the beautiful Spix's macaw, so the records are revealing a worrying new trend: 'For the first time,' the report indicates, 'mainland extinctions are outpacing island extinctions.'

Extinction in itself is not alarming. More species are extinct than are living today, but what worries scientists is the rate of extinctions they're witnessing. Life has certainly had its ups and downs. Aside from the five major mass extinctions in the past, when life almost came to a sticky end, there have been many smaller events, and there is a constant natural background rate of extinctions – about 0.1 extinctions per million species per year or one extinction every 10 years – but today the rate could be up to 1,000 times greater.

A Royal Botanic Gardens and Stockholm University study can be more precise about recent plant extinctions. They reveal that 571 species have disappeared during the past 250 years, thought to be 500 times the background rate; but it is not all bad news. They also found that half of all orchid species, which had been declared extinct, have been rediscovered, and are alive and well.

Nevertheless, it is estimated that a future extinction rate could be as high as 10,000 times the background rate, and that up to 50 per cent of all living species could be extinct by the middle of the twenty-first century. Scientists are calling it the 'sixth mass extinction'; one caused not by space rocks or volcanic eruptions, but by human activities. Our impact on the planet is such that some have also proposed a new epoch in Earth's history – the Anthropocene. It's not a welcome accolade, but a sombre indictment of the way we have mismanaged and continue to mismanage the seven amazing worlds on our one extraordinary planet.

Opposite
The surviving two female northern white rhinos are so precious that armed rangers guard them. With the last male dead, scientists pin their hopes on IVF. The plan is to use stored sperm from long-deceased males and eggs from the surviving females. In this way, the subspecies might be saved.

Index

Acknowledgements

Neither the book nor the TV series would exist without the generous help of scientists, wildlife researchers, conservation & wildlife organisations, universities and local wildlife experts. Ultimately, it is these people and organisations that we rely on so much to make these shows.

For turning these stories into compelling sequences we would like to thank our production team; for the past four years working on *Seven Worlds* has been their one and only world. Thanks for your dedication, passion and hard work.

Behind the cameras and behind the scenes are the camera crews, sound recordists, editors, edit assistants and technicians, who together make these stories look and sound so good. Thank you for continuing to raise the bar.

And to all the friends and families, on behalf of everybody, we'd like to thank you for your patience, understanding and support. It is more important than you'll ever know.

Finally, for this stunning book that complements the series so well, we'd like to thank Michael Bright and all the team at Penguin Random House.

Scott Alexander & Jonny Keeling

PRODUCTION TEAM
Sir David Attenborough

Charlotte Moore
Tom McDonald

Abigail Lees
Adam Oldroyd
Aleks Grant
Alex Page
Angel Garcia-Rojo
Caroline Cox
Chadden Hunter
Charlotte Bostock
Claire Thompson
Cleone Fox
Craig Haywood
Daniel Butt
Deya Ward
Elizabeth White
Emma Hatherley
Emma Napper
Felicity Lanchester
Fredi Devas
Gemma Templar
Giles Badger
Greg Slater
Jack Delf
Jane Atkins
Jess Webster
Jo Avery
Jo Haley
Jo Stead
Joel Rogers
Jonny Keeling
Julian Hector
Kate Gorst
Katie Hall
Kelsie Chappell
Kiri Cashell

Lauren Jackson
Lucy Wells
Maddie Close
Maria Norman
Mary Melville
Michael Becker
Nardine Groch
Natasha Prymak
Nick Green
Nicola Kowalski
Patrick Evans
Rosie Gloyns
Sadie Coles
Sarah Whalley
Scott Alexander
Sophie Lanfear
Sue Luton
Theo Webb
Tom Parry
Vicky Knight
Zoë Beresford

**CAMERA AND
SOUND TEAM**
Alastair MacEwen
Alastair Smith
Alek Kydd
Alex Vail
Andre Rerekura Creative
Andrew Thompson
Barny Trevelyan Johnson
Barrie Britton
Benjamin Cunningham
Bernt Bruns
Bertie Gregory
Cameron Board
Charles Davis
Chris Watson
Christopher Tangey
Cristian Dimitrius

Dan Beecham
Daniel Hunter
Darryl MacDonald
David Herasimtschuk
David Parer
Didier Noirot
Edward Saltau
Erin Ranney
Espen Rekdal
Ewan Donnachie
Gavin Newman
Gemilang Dini Ar-Rasyid
Grant Baldwin
Guillermo Armero
Hector Skevington-Postles
Heliguy.tv
Henry Mix
Howard Bourne
Hugh Miller
Iqbal Sunni
Jack Hynes
Jacky Poon
James Ewen
James Loudon
Jamie McPherson
Jeremy Monroe
Jesse Wilkinson
João Paulo Krajewski
John Aitchison
John Brown
John Shier
John Totterdell
Jonathan Jones
Joshua Jorgensen
Julie Moniere
Justin Hofman
Justin Maguire
Justine Evans
Kieran O'Donovan
Kopterworx

Li Shuai
Lindsay McCrae
Mark MacEwen
Mark Payne-Gill
Mark Smith
Martyn Colbeck
Max Hug Williams
Michael Patrick O'Neill
Neil Anderson
Nick Turner
Oliver Jelley
Olly Scholey
Paul Klaver
Pete McCowen
Robert Hawthorne
Roger Munns
Rolf Steinmann
Russel Laman
Russell Maclaughlin
Sandesh Kadur
Santiago Cabral
Scott Snider
SeaLife Differently
Sean Scott
Shane Moore
Simon De Glanville
Sophie Darlington
Ted Giffords
Tim Laman
Toby Strong
Tom Chapman
Tom Crowley
Tom Fitz
Tom Walker
Tony Driver

POST PRODUCTION
Films at 59
Miles Hall
Wounded Buffalo

MUSIC
Bleeding Fingers
Hans Zimmer

Andrew Christie
Christopher Braide
Christopher King
David Russell
Jacob Shea
Natasha Pullin
Nick Baxter
Russell Emanuel
Sia
Steve Kofsky

FILM EDITORS
Andy Netley
Angela Maddick
Dave Pearce
Jack Roberts
Matt Meech
Nigel Buck

Doug Main
Emma Jones
Nick Carline
Owen Porter
Robbie Garbutt

ONLINE EDITORS
Franz Ketterer
Wesley Hibberd

DUBBING EDITORS
Kate Hopkins
Tim Owens

DUBBING MIXERS
Chris Domaille
Graham Wild

COLOURIST
Adam Inglis

GRAPHIC DESIGN
Hello Charlie

BBC STUDIOS SALES & DISTRIBUTION
Mark Reynolds
Patricia Fearnley

Hayley Moore
Monica Hayes

CO-PRODUCERS
BBC America
China Media Group CCTV9
France Télévisions
Tencent Penguin Pictures
ZDF

SCIENTIFIC CONSULTANT
Professor Iain Stewart

TECHNICAL CONSULTANTS
Colin Jackson
Gordon Leicester

SERIES DEVELOPMENT
Dan Huertas
Doug Hope
Emily Miller
Jonny Keeling
Luke Ward
Nick Easton
Renee Godfrey

WITH THANKS TO
Alex Board
Andrew Downey
Christopher Scotese
Copernicus Sentinel
Donna Gomes
Glastonbury Festival
Jemal Guerrero
Jo Hall
NASA
Nathan Garofalos
Ol Pejeta Conservancy
Dr Patrick Avery
Dr Paul Reavley
Susan Attenborough
Truenorth (Iceland)

ASIA
Altay Films
Amod Zambre
Anatoly Kochnev
Anna Smirnova
Bromo Tengger Semeru National Park
Dr Bariushaa Munkhtsog
Dr Cheryl Knott
Dr Zulfi Arsan
Elena Bulatova
Ellen Xu

Evgeniya Saevich and staff of Kinross
Explore Kamchatka
Fan Penglai
Felis Creations
Harry Amies & Amir Rezazadeh
Harshal Bhosale
indoXPLORE
International Rhino Foundation
Joe Faithfull
Khustain Nuruu National Park
Kronotsky State Nature Biosphere Reserve
Lahuka
Martha Madsen
Maxim Chakilev
Maxim Deminov
Nomads Expeditions
Pam Fogg
Rory May
Scuba Zoo
Shennongjia National Nature Reserve
Tanjung Gunung National Park
Tatyana Minenko
Teluk Cenderawasih National Park
The Gecko Project
The Lighthouse Consultancy
Tim Fogg
Valeriy Kalyarakhtyn
Viktoria Belger
Way Kambas National Park
Yayasan Badak Indonesia

AUSTRALASIA
Acknowledgment of Country
Alex Humphries
Anangu (Uluru) Aboriginal Community
Andrew Kaineder
Barrington Tops National Park
Ben Stevenson
Brad Purcell
Brett McNamara
Cooper Creek Wilderness
Curtin Springs
Darroch Donald
David Holman
David Lornie
Elsey National Park
Emily Belton
Eungella National Park
Fowlers Gap Arid Zone Research Station
Heli Surveys
Intomedia
Jason Cummings
Journey Beyond Rail Expeditions
Jürgen Otto
Kosciuszko National Park

Luke Hasaart
Mardudhunera Aboriginal Community
Minibeast Wildlife
Mulligans Flat Woodlands Sanctuary, Woodlands and Wetlands Trust
Mungo National Park, Willandra Lakes World Heritage Centre
Murray Kille
Murujuga Aboriginal Community
Mutthi Mutthi Aboriginal Community
Namadgi National Park
Ngyiampaa Aboriginal Community
Nick Mooney
Ningaloo Marine Park
Noonbah Station
One Ocean International
Paakantji Aboriginal Community
Rainforest Scuba
Rex Neindorf
Steve Baldwin
Tasmania Parks and Wildlife Service
The Thylacine Museum
Thorntonia Staion
Tkoda Hewett
Uluru-Kata Tjuta National Park
Wendy & Alan Page
Wildlife Habitat Port Douglas
Wong-Goo-Tt-Oo Aboriginal Community
Yaburara Aboriginal Community
Yinjibarndi Aboriginal Community

EUROPE
Alessandro Di Federico from Elandra Productions srl
Autoritatea Aeronautică Civilă Română
Bio Aqua Pro kft
Bruno D'Amicis
Civil Aviation, HM Government of Gibraltar
Danube Delta Bisophere Reserve
Dr Béla Kiss
Dr Eric Shaw
Friedhöfe Wien GmbH
German Garrote
Julian Rad
Junta de Andalucia, Consejeria de Agricultura, Pesca y Medio Ambiente
Katarina Kanduč
Kristóf Málnás
Knowledge, Vision and Talent S.L
Life+Iberlince

Martinselkonen Wilderness Centre
Miguel Simon Mata
National Park of Abruzzo, Lazio and Molise
Oppdal Safari
Oulanka National Park
Postojnska jama d.d.
Primož Gnezda

ANTARCTICA
Akademik Ioffe
Alfred-Wegener Institut Helmholtz-Zentrum für Polar und Meeresforschung (AWI)
Amy Moran
Andres Barbosa Alcon
Antonio Quesada
Ari Friedlaender
Arnaud Tarroux
Athena Dinar
Bettina Meyer
Bill Morris
Bird Island Research Station
Bree Texter
Britney Schmidt
British Antarctic Survey
Catrin Thomas
Charles Triggs
Charly Bainbridge
Claudia & Jurgen Kirchberger
Coli Whewell
Comité Polar Español
Dave Roberts
David Johnston
Department Of Conservation, New Zealand
Derren Fox
Diego Fernando Naselli
Dirección Nacional del Antártico, Instituto Antártico Argentino
Edd Hewett
Ejército de Tierra, Base Gabriel de Castilla
Elaine Hood
Emma Carroll
Estación Primavera, Cierva Cove
Eva Leunissen
Evan Kenton
Evie Rainey
Fernanda Millicay
Gary Paget
Gene Butler
Government of South Georgia and the South Sandwich Islands
Harald Magnus Eilertsen
Harold Jäger
Helena Herr
James B. McClintock
Jamie Coleman
Jan Kendzia
Jay Rotella
Jen Jackson

Johan Bondi
John Bowles
John Dickens
John Janssen
Jon Hugo Strømseng
Kirsten Neuschafer
Ponant and the crew of Le Boreal
Mark Shortt
Marta Guerra
Mats-Ola Finn
McMurdo Station
Mercedes Santos
Michael Dinn
Morten Iverson
NHNZ
Niko Dubreuil
NMFS MMPA Permit No. 21486
Norwegian Polar Institute, Norwegian Antarctic Research Expedition 2017-2018
NSF – National Science Foundation
Osama Mustafa
Otago University
Otto Puolakka
Paul Cziko
Peter Sammonds
Skip Novak (Pelagic Expeditions)
Phil Trathan
Eleanor Floyd
Stuart Doubleday
Polar Regions Department, Foreign and Commonwealth Office
All Crew and Scientists on Polarstern Expedition PS112
Richard Phillips
Rob Robbins
Robert Whitworth
Rodolfo Sánchez
Sacha Viquerat
Wolf Kloss & Jeannete Talavera (SIM Expeditions)
Sebastien Descamps
Simon Van Dam
Stephen Bradley
Steve Brown
Steve Dawson
Steve Little
Steve Rupp
Sven Lidström
Tom Brough
United States Antarctic Program
Tom Schwartz
University College London, Earth Sciences
Valentin Beneitez
Valentine Kass
Weddell Seal Project, NMFS Nos. 1032-1917 & 21158
Will Rayment

NORTH AMERICA
Annie Band & Jon Hunt
Bryan Thomas
Casper Cox
Connor Stefanison
Cora Berchem
Craig Littauer
Crystal River National Wildlife Refuge, Crystal River, Florida courtesy of the U.S. Fish and Wildlife Service
Daniel Batchelor
Daniel Grayson
Darrell Thomas
David Dearth
Dawson Dunning
Eric Welscher-Bilodeau
Flores Island Park, British Columbia, Canada
Greg Bogdan
Hilary Webber
Howard Leftwich
Issac Szabo
Jackie Stearns
James Green
Jamie Larson
Jared Hires
John Forde
John Spann
Joyce Kleen
Laura Hughes
Libby & Paul Hartfield
Lynn Faust
Mario Martinez
Mark Emery
Mark Zloba
Marni Walsh
Matthew James Allen
Monument Valley Navajo Tribal Park
Nerissa Okiye & Sally Beynon
Phoebe Fitz
Quent Plett
Ryan Durack
Sam & Cam Eddy
Save the Manatee Club
Seal River Lodge
Terinda Whisenant
The Navajo Nation Film Office
Wayne Hartley
Zoe Rossman

SOUTH AMERICA
Alfredo Enrique Martinez Olivares
Ana Arroyo
Caesar Lopez
Claudio Bustos
Coli Whewell
Corporación Nacional Forestal
Diego Araya
Donald J. Brightsmith
Dr Anne Savage
Emilio White
Felix Segundo Medina Carrasca
Fernando Ayala-Varela
Fernando Blanco
Fernando Mateos Gonzalez
Freddy Vergara
Fundación Proyecto Tití
George Olah
Guillermo Knell
Harly Garcia
Iguaçu National Park - Federal Conservation Unit, managed by ICMBio
Chico Mendes de Conservação da Biodiversidade
Irene Alejandra Ramírez Merida & Ximena Alejandra Álvarez Bustos
Jesus Vergara
Johanna Vega
Jose Luis Escandela
Justin Yeager
Lucas Salter
Manuel Sanchez
Manuel Panaijo
Marcial Urbina
Marie Castro
Mark Pepper
Nancy Olivares
Oliver Laker
Omar Torres Carvajal
Pastora Donoso
Rainforest Expeditions
Rebeca Justicia
Roberto Donoso
Rocío Guzmán
Rosamira Guillen
Santiago Molina
SERNANP – Servicio Nacional de Areas Naturales Protegidas por el Estado
Stefano Raffo Porcari & Luis Felipe Raffo
Stuart Scuba
Susana Cárdenas Alayza
The Macaws Project
Torres del Paine National Park, Chile
Vicente Montero

AFRICA
ANSSO
Czech Academy of Sciences, Institute of Vertebrate Biology
Department Of National Parks and Wildlife, Zambia
Dr Abdeljebbar Qninba
Dr Julia Riley
Dr Kenneth McKaye
Dr Liran Samuni
Dr Martin Reichard
Dr Roman Wittig
Dr Tony Phelps
Ebrin Brou
Etty Varley
Femke Broekhuis
HEEED Malawi
Ingrid Wiesel / Brown Hyena Research Project
Ivonne Kienast
James Kimaru
Kenya Wildlife Trust
Kolmanskop Tour Company
Lake Bogoria National Reserve
Le Ministère de l'Economie Forestière, du Développement Durable et de l'Environnement, Congo
Mana Pools National Park
Mara Cheetah Project
Matej Polačik
Mike Saunders
Milou Groenenberg
Ministry of Communication, Malawi
Ministry of Environment and Tourism, Namibia
Ministry of Mines and Energy, Namibia
NamDeb
Namibia Film Commission
Namibian Civil Aviation Authority
Namibian Film Commission
Narok County, Kenya
Nick Murray
Office of Ivory Coast Parks and Reserves
Radim Blazek
Robin Pope Safaris
Samuel Munene
Sandstone film logistics, Namibia
South African National Parks or Augrabies Falls National Park
Stanley Kinyolo
Tai Chimpanzee Project
Tswalu Kalahari Reserve
Viewfinders Kenya Ltd
Wendy Panaino
Wildlife Conservation Society
Wim Vorster
Zanne Labuschagne
Zimbabwe Parks and Wildlife Management Authority

10 9 8 7 6 5 4 3 2 1

BBC Books, an imprint of Ebury Publishing
20 Vauxhall Bridge Road,
London SW1V 2SA

BBC Books is part of the Penguin Random House group of companies whose addresses can be found at global.penguinrandomhouse.com

Penguin
Random House
UK

This book is published to accompany the television series entitled *Seven Worlds, One Planet,* first broadcast on BBC One in 2019.

Executive producer: Jonny Keeling
Series producer: Scott Alexander

First published by BBC Books in 2019
www.penguin.co.uk

A CIP catalogue record for this book is available from the British Library

978-1-785-94412-3

Commissioning Editor: Albert DePetrillo
Project Editor: Nell Warner
Picture Research: Laura Barwick
Image Grading: Stephen Johnson, www.copyrightimage.co.uk
Design: Bobby Birchall, Bobby&Co
Production: Rebecca Jones

Printed and bound in Italy by Printer Trento

Penguin Random House is committed to a sustainable future for our business, our readers and our planet. This book is made from Forest Stewardship Council® certified paper.

MIX
Paper from responsible sources
FSC® C018179

Picture credits